SURVEYS IN NUMBER THEORY

Developments in Mathematics

VOLUME 17

Series Editor:
Krishnaswami Alladi, *University of Florida, U.S.A.*

SURVEYS IN NUMBER THEORY

Edited by

KRISHNASWAMI ALLADI
University of Florida, U.S.A.

 Springer

Krishnaswami Alladi
Department of Mathematics
University of Florida
Gainesville, FL
U.S.A.
alladi@math.ufl.edu

ISSN:1389-2177
ISBN-13: 978-1-4419-2689-0   e-ISBN-13: 978-0-387-78510-3
DOI: 10.1007/978-0-387-78510-3

Mathematics Subject Classification (2000): Primary: 01-02, 05A17, 15A15, 11A, 11A07, 11A25, 11A41, 11B05, 11B13, 11B34, 11B65, 30C, 33C10, 11G05, 11J89, 11P82; Secondary: 01A60, 05A15, 05A30, 11F12, 11F20, 11P81, 11P83

Printed on acid-free paper.

springer.com

Contents

Preface

During the academic year 2004–2005, the Mathematics Department of the University of Florida conducted a *Special Year in Number Theory and Combinatorics*. This was the fourth of a six-year program of special years, focusing on various areas of strength and tradition in the department. Each special year featured international conferences, training workshops, lectures during the year by visitors, and survey lectures of wide appeal. The number theory part of the program in 2004–2005 was highlighted by an International Conference on Additive Number Theory in November 2004, the Seventh Erdős Colloquium by Fields medalist Borcherds, and several history lectures by very eminent mathematicians throughout the year. These history lectures traced major developments in various important aspects of number theory.

The number theory part of the program, including the conference on additive number theory, was organized by Krishnaswami Alladi, Alexander Berkovich, Frank Garvan, and Hamza Yesilyurt. The program received external funding from the National Science Foundation, the National Security Agency, and the Number Theory Foundation. Internal support was provided by the Department of Mathematics, the College of Liberal Arts and Sciences, and the Office of Research and Graduate Programs of the University of Florida. My thanks to the organizers of the program and all the agencies for their support.

This volume consists of a selection of featured talks given either at the conference or as history lectures during the year. Number theory has a wealth of long-standing problems, the study of which over the years has led to major developments in many areas of mathematics. The papers assembled here span a broad range of areas in number theory and survey important developments tied to current advances. The surveys are written by leaders in the discipline: George Andrews, Bruce Berndt, David Bressoud, Hershel Farkas, Mel Nathanson, Ken Ono, and Michel Waldschmidt. The topics covered reflect the wide range of their expertise and include the theory of partitions and q-hypergeometric series, mock theta functions, modular forms, identities from Ramanujan's lost notebook, theta functions, elliptic functions, alternating sign matrices, and the Weyl character formula, transcendental number theory, and additive number theory.

The theory of partitions and q-hypergeometric series was founded by Euler in the mid-eighteenth century. The subject underwent a glorious transformation in the early twentieth century under the magic touch of the Indian genius Srinivasa Ramanujan, who discovered remarkable identities that provided surprising connections between seemingly unrelated areas of mathematics. The theory of partitions and q-series is now at the crossroads of number theory, combinatorics, analysis, and the theory of modular forms, and has implications in physics and computer science.

One of the most dramatic developments in the past few years has been the work of Ken Ono and Kathrin Bringmann, connecting Ramanujan's mock theta functions with Maass wave forms. Many consider the mock theta functions to be among Ramanujan's deepest contributions. Shortly before his death in 1920, Ramanujan communicated his discovery of the mock theta functions in a letter to G. H. Hardy, of Cambridge University. These mock theta functions possessed transformations that enabled one to evaluate their coefficients with great precision, similar to those of certain partition-generating functions that had representations in terms of theta functions. Yet the exact connections between mock theta functions and the theory of modular forms or theta functions remained a mystery. The recent work of Ono and Bringmann is the key that unlocks this mystery. Several important aspects of this progress are described by Ono in his paper.

Our understanding of Ramanujan's discoveries has been greatly enhanced in the last few decades due to the work of several leading mathematicians, most notably George Andrews and Bruce Berndt. Ramanujan's original two notebooks have been edited in a five-volume series (Springer) by Berndt, who has explained almost all of the formulas in those notebooks. And now George Andrews and Bruce Berndt are editing Ramanujan's "lost notebook" in a similar fashion. There are still several aspects of Ramanujan's identities in his original two notebooks and in the lost notebook that merit a closer study, because, as it often turns out, when one discovers important identities and formulas at the interface of classical analysis and number theory, they can be traced to the writings of Ramanujan! The paper of Bruce Berndt, Yoongbok Lee, and Jaebum Sohn is about identities in Ramanujan's lost notebook that yield the fundamental formulas of Guinand and Koshliakov concerning functional equations of nonanalytic Eisenstein series which are now central in the theory of Maass wave forms. Indeed, Ramanujan gives several new applications of Guinand's formula as noted by Berndt, Lee, and Sohn in their paper.

The theory of partitions has continued to be a vibrant area of research because partition identities have arisen in surprising ways in a variety of settings, such as the celebrated Rogers–Ramanujan identities in the study of vertex operators of Lie algebras, and in the investigation of models in conformal field theory and statistical mechanics. George Andrews, the leading authority on partitions in the last half century, has been intimately connected with several of these major developments. In his survey paper, Andrews discusses multipartitions, which have proved useful in the theory of Lie algebras. He builds on the earlier important work of Oliver Atkin to provide congruences for multipartitions, and explains how multipartitions are connected to the tripentagonal number theorem. He also discusses connections with the Rogers–Ramanujan identities and Ramanujan's mock theta functions.

Theta functions are among the most fundamental objects in mathematics. They arise as the solution to the heat equation in physics, can be used to yield the functional equation of the Riemann zeta function, and play a crucial role in the study of partition-generating functions and other generating functions involving quadratic forms. Theta functions also are crucial in the study of Riemann surfaces. Hershel Farkas, an authority on Riemann surfaces, discusses the role of theta functions in analysis and number theory, and demonstrates how the relations satisfied by theta functions can be used to yield a variety of identities in additive number theory involving sums of squares, partitions, and multiplicative functions.

Closely related to theta functions are elliptic functions, which have played, and continue to play, a central role in several areas of mathematics, especially number theory. Transcendental number theory is a fascinating subject. One reason is that although "almost all" numbers are transcendental, it is very difficult to determine whether a given number is transcendental. The transcendence of certain classical constants, or special values of fundamental functions, have been investigated closely, but there is still much to be learned. One of the fundamental results due to Siegel is that if the two invariants of the Weierstrass elliptic function are algebraic numbers, then at least one of the two periods of the elliptic function must be transcendental. Michel Waldschmidt provides a comprehensive survey of elliptic functions and transcendence. He first describes the landmark results in transcendental number theory relating to the exponential and logarithm, and then discusses in detail the methods and results involving the transcendence of special values of the elliptic functions and constants related to them.

Additive number theory is a very broad area within which lies the theory of partitions, and the subject of representation of integers as sums of squares, cubes, etc. There is a vast literature in additive number theory on problems of the following type: Given a subset **A** of the positive integers, when does it form a basis for the integers? That is, when can every positive integer be represented as the sum of at most a (fixed) bounded number of elements of **A**? There is also the question of enumerating the number of representations of integers as sums of members coming from a given subset **A**. And then there is the "inverse problem"; namely, if we are given a nonnegative arithmetic function, when can the values of this function be the number of representations of integers from some subset **A** of the integers? Mel Nathanson provides a survey of such inverse problems in additive number theory.

Finally, the paper of David Bressoud provides a charming survey of certain important determinant evaluations, starting with the rudimentary Vandermonde, and ending with the Izergin–Korepin determinant expansion for the six-vertex model in statistical mechanics. En route, he discusses the alternating-sign matrix conjecture, and the ways in which this was proved.

Thus this volume provides a kaleidoscopic view of number theory and contains informative summaries of several important aspects of the subject. Special thanks to

Ann Kostant, of Springer, for publishing these selected survey lectures, and to Frank
Garvan for help in preparing the volume according to Springer's specifications.

Krishnaswami Alladi
Gainesville, Florida
March, 2008

1

A Survey of Multipartitions: Congruences and Identities

George E. Andrews

The Pennsylvania State University, University Park, PA 16802, USA
andrews@math.psu.edu

Summary. The concept of a multipartition of a number, which has proved so useful in the study of Lie algebras, is studied for its own intrinsic interest. Following up on the work of Atkin, we shall present an infinite family of congruences for $P_k(n)$, the number of k-component multipartitions of n. We shall also examine the enigmatic tripentagonal number theorem and show that it implies a theorem about tripartitions. Building on this latter observation, we examine a variety of multipartition identities connecting them with mock theta functions and the Rogers–Ramanujan identities.

Key words: Partitions, multipartitions, partition congruences, tripentagonal number theorem.

2000 *Mathematics Subject Classifications*: Primary 05A17; Secondary 11P81, 11P83, 05A30

1 Introduction

In 1882, J. J. Sylvester [28] broke new ground in the theory of partitions. Throughout most of the nineteenth century, partitions of integers were viewed primarily as an auxiliary aid in the theory of invariants. Sylvester's monumental paper [28] revealed that partitions were themselves interesting mathematical objects with a surprising rich arithmetic/combinatorial theory.

Today we find multipartitions in an analogous situation. We use M. Fayers's definition of a multipartition [16, p. 4].

A *multipartition of n with r components* is an r-tuple $\lambda = (\lambda^{(1)}, \ldots, \lambda^{(r)})$ of partitions such that $|\lambda^{(1)}| + \cdots + |\lambda^{(r)}| = n$. If r is understood, we shall just call this a multipartition of n. As with partitions, we write the unique multipartition of 0 as $\emptyset$, and if λ is a multipartition of n then we write $|\lambda| = n$.

Partially supported by National Science Foundation Grant DMS 0457003.

K. Alladi (ed.), *Surveys in Number Theory*, DOI: 10.1007/978-0-387-78510-3_1,

There are indeed various scattered results about multipartitions. In many papers on the representation theory of Lie algebras (e.g., [13], [16]), some with applications to physics [12], we see multipartitions playing an important auxiliary role. However, there has not previously been a survey devoted to their intrinsic interest.

It is not the case that the generating functions for multipartitions have not been examined previously. Indeed, Atkin [10], Cheema et al. [14], Gupta et al. [20], [21] and others have examined congruences and other arithmetic properties of $P_k(n)$, where

$$\sum_{n=0}^{\infty} P_k(n)q^n = \prod_{n=1}^{\infty} (1 - q^n)^{-k}. \tag{1.1}$$

We see immediately by Euler's standard argument [3, Chapter 1] that $P_k(n)$ is the number of k-component multipartitions of n. Atkin [10] has supplied the most extensive list of congruences for $P_k(n)$, building on his success in proving Ramanujan's "11^n" conjecture [9]. In Section 2, we prove the following.

Theorem 1. *For every prime $p > 3$, there are $(p + 1)/2$ values of b in the interval $[1, p]$ for which*

$$P_{p-3}(pn + b) \equiv 0 \qquad (\text{mod } p)$$

for all $n \geq 0$.

When $p = 5$, the b values are 2, 3, 4; for $p = 7$, they are 2, 4, 5, 6; and for $p = 11$, they are 2, 4, 5, 7, 8, 9.

The proof of Theorem 1 requires only the original elementary method of Ramanujan as extended in [8].

Section 3 leads us to the starting point of our survey. Namely, can the combinatorial significance of the tripentagonal number theorem [6, (1.3)] be placed in the world of multipartitions? That is, do we have a combinatorial interpretation for

$$\sum_{i,j,k \geq 0} \frac{q^{i^2+j^2+k^2}}{(q;q)_{i+j-k}(q;q)_{i+k-j}(q;q)_{j+k-i}}$$

$$= \frac{\sum_{n,m,p=-\infty}^{\infty}(-1)^{n+m+p}q^{n(3n-1)/2+m(3m-1)/2+p(3p-1)/2+nm+np+mp}}{\prod_{n=1}^{\infty}(1 - q^n)^3}. \tag{1.2}$$

We use the standard notation [19, (1.2.5)]

$$(a;q)_n = (a)_n = \prod_{j=0}^{n-1}(1 - aq^j),$$

and for future reference [19, (1.2.4)],

$$(a_1, a_2, \ldots, a_m; q)_n = (a_1; q)_n(a_2; q)_n \cdots (a_m; q)_n.$$

This leads to a vast array of multipartition identities, which are further explored in Section 4. Section 5 examines the relationship of several generalizations of classical partitions to multipartitions, and Sections 6 and 7 conclude the survey with a brief account of the further possibilities for multipartitions.

2 Congruences

The main object of this section is to prove Theorem 1, and by so doing, to suggest that this is the tip of a congruential iceberg. To prove Theorem 1 we must recall a slightly weakened version of the main result in [8].

Theorem 2 (Extended Ramanujan Congruence Theorem). *Let $p > 3$ be a prime. Let $0 < a < p$ and b be integers with $-a$ a quadratic nonresidue mod p. Let $\{\alpha_n\}_{n=-\infty}^{\infty}$ be any doubly infinite sequence of integers. Then there exists an integer $c = c_p(a, b)$ such that the coefficient of q^{pN} for any $N > 0$ in*

$$\frac{\sum_{n=-\infty}^{\infty} \alpha_n q^{a\binom{n}{2}+bn+c}}{(q;q)_{\infty}^{p-3}}$$

is divisible by p. The number c is $\overline{8}(a(2b\overline{a} - 1)^2 + 1)$ modulo p, and $\overline{m}$ denotes the multiplicative inverse of m modulo p.

Proof of Theorem 1. To deduce Theorem 1, we note that for any specific n, say $n = \nu$, we may select the $\{\alpha_n\}_{n=-\infty}^{\infty}$ by $\alpha_n = 0$ if $n \neq \nu$, $\alpha_\nu = 1$. Thus the coefficient of q^{pN} in

$$\frac{q^{a\binom{\nu}{2}+b\nu+c}}{(q;q)_{\infty}^{p-3}} = \sum_{n\geq 0} P_{p-3}\left(n - a\binom{\nu}{2} - b\nu - c\right) q^n$$

is divisible by p. So the proof of Theorem 1 is reduced to showing that the exponents $a\binom{n}{2} + bn + c$ run over $(p+1)/2$ different residue classes modulo p. To this end we consider the congruence

$$a\binom{n}{2} + bn + c \equiv r \pmod{p}.$$

This congruence is equivalent to

$$(n + \overline{2}(\overline{a}b - 1))^2 \equiv 2\overline{a}r - \overline{4}(\overline{a}b - 1)^2 \pmod{p}.$$

Now as n runs over a complete residue system modulo p, the left side of this last congruence takes the value of each of the $(p-1)/2$ quadratic residues twice and 0 once for a total of $(p+1)/2$ different values modulo p. For each of these values there is a unique value of r, which is what was necessary to prove. □

One of the most appealing aspects of the theory of congruences for $p(n)$ $(= P_1(n))$ concerns the "crank." The crank of a partition is defined as follows:

Definition. For a partition π, let $l(\pi)$ the largest part of π, $\omega(\pi)$ the number of ones in π, and $\mu(\pi)$ the number of parts of π larger than $\omega(\pi)$. The crank $c(\pi)$ is given by

$$c(\pi) = \begin{cases} l(\pi) & \text{if } \omega(\pi) = 0, \\ \mu(\pi) - \omega(\pi) & \text{if } \omega(\pi) > 0. \end{cases}$$

In [7], it is shown that the generating function for $c(m, n)$, the number of partitions of n with crank m, is given by

$$\sum_{n \geq 0} \sum_{m=-\infty}^{\infty} c(m, n) z^m q^n = \frac{(q; q)_\infty}{(zq, z^{-1}q; q)_\infty},$$

except for $n = 1$.

The results in [17] and [7] together reveal the following combinatorial interpretation of the Ramanujan congruence

$$p(5n + 4) \equiv 0 \pmod{5}.$$

Namely, if $C(j, M, n)$ denotes the number of partitions of n with crank $\equiv j$ (mod M), then for $0 \leq j \leq 4$,

$$C(j, 5, 5n + 4) = \frac{1}{5} p(5n + 4).$$

We now define

$$\sum_{n \geq 0} \sum_{m=-\infty}^{\infty} b(m, n) z^m q^n = \frac{(q; q)_\infty^2}{(zq, z^{-1}q; q)_\infty^2}$$

and

$$B(j, M, n) = \sum_{m=-\infty}^{\infty} b(Mm + j, n).$$

It is immediate that

$$\sum_{m=-\infty}^{\infty} b(m, n) = P_2(n),$$

and thus Theorem 3 below provides us with a new proof that $5 \mid P_2(5n + 3)$.

It is, of course, possible to provide an interpretation for $b(m, n)$ in terms of modified bipartitions. Namely, one has to allow three copies of 1 to be used: 1_{-1}, 1_0, 1_1 with the added condition that 1_0 is not repeated in either component and if it appears in one component and not the other then that bipartition is counted with weight -1. This then provides a combinatorial setting for the fact that $c(1, 1) = 1$, $c(0, 1) = -1$, and $c(-1, 1) = 1$.

Theorem 3. *For $0 \leq j \leq 4$,*

$$B(j, 5, 5n + 3) = \frac{1}{5} P_2(5n + 3).$$

Proof. Garvan [17, (1.30) and Section 3] proved the following formula from Ramanujan's lost notebook [25, p. 20]:

$$\frac{(q;q)_\infty}{(\xi q;q)_\infty(\xi^{-1}q;q)_\infty}$$

$$= A(q^5) - q(\xi+\xi^{-1})^2 B(q^5) + q^2(\xi^2+\xi^{-2})C(q^5) - q^3(\xi+\xi^{-1})D(q^5),$$

where ξ is any primitive fifth root of unity, and

$$A(q) = \frac{(q^5;q^5)_\infty G(q)^2}{H(q)},$$

$$B(q) = (q^5;q^5)_\infty G(q),$$

$$C(q) = (q^5;q^5)_\infty H(q),$$

$$D(q) = \frac{(q^5;q^5)_\infty H(q)^2}{G(q)},$$

$$G(q) = \frac{1}{(q,q^4;q^5)_\infty},$$

$$H(q) = \frac{1}{(q^2,q^3;q^5)_\infty}.$$

We note for future reference that

$$A(q)D(q) = B(q)C(q). \tag{2.1}$$

Hence

$$\sum_{j=0}^{4}\sum_{n=0}^{\infty} B(j,5,n)\xi^j q^n = \left(\frac{(q;q)_\infty}{(\xi q,\xi^{-1}q;q)_\infty}\right)^2$$

$$= (A(q^5) - q(\xi+\xi^{-1})^2 B(q^5) + q^2(\xi^2+\xi^{-2})C(q^5) - q^3(\xi+\xi^{-1})D(q^5))^2.$$

Therefore

$$\sum_{j=0}^{4}\sum_{n=0}^{\infty} B(j,5,5n+3)\xi^j q^{5n+3}$$

$$= -2q^3((\xi+\xi^{-1})A(q^5)D(q^5) + (\xi+\xi^{-1})^2(\xi^2+\xi^{-2})B(q^5)C(q^5))$$

$$= -2q^3 A(q^5)D(q^5)(\xi+\xi^{-1})(1 + (\xi+\xi^{-1})(\xi^2+\xi^{-2})) \quad \text{(by (2.1))}$$

$$= -2q^3 A(q^5)D(q^5)(\xi+\xi^{-1})(1 + \xi^3 + \xi^4 + \xi + \xi^2)$$

$$= 0.$$

Consequently, all of the $B(j, 5, 5n+3)$ must be identical in value, since otherwise, we would have constructed a new polynomial of degree ≤ 4 satisfied by ξ different from $1 + \xi + \xi^2 + \xi^3 + \xi^4$, which is *the* irreducible polynomial of which ξ is a root, a contradiction.

Finally, since all the $B(j, 5, 5n+3)$ are equal, each must be equal to $P_2(5n+3)/5$.

$\square$

Frank Garvan [18] has independently extended these methods in a forthcoming paper that goes well beyond what has been suggested here. It should also be noted that K. Mahlburg [23] has proved amazing congruence theorems for the crank, and this suggests that it may be possible to do much more with the bicrank and further extensions of the crank to multipartitions.

3 The Triple Theta Series

We devote this section to

$$S_3(q) = \frac{\sum_{n_1,n_2,n_3=-\infty}^{\infty} (-1)^{n_1+n_2+n_3} q^{n_1^2+n_2^2+n_3^2+\binom{n_1+n_2+n_3}{2}}}{(q)_\infty^3}, \tag{3.1}$$

the theta series side of (1.2). Our object will be to obtain as succinct as possible representation of $S_3(q)$ as a linear combination of infinite products.

Theorem 4.

$$S_3(-q) = \frac{1}{(-q; -q)_\infty (q^2; q^2)_\infty} \sum_{n,j=-\infty}^{\infty} (-1)^j q^{\binom{n}{2}+n^2-4nj+6j^2}. \tag{3.2}$$

Proof. Using the notation

$$[z^m] \sum_{n=0}^{\infty} a_n z^n = a_m, \tag{3.3}$$

we see that

$$S_3(q) = [z^0] \frac{\left(\sum_{n=-\infty}^{\infty} (-z)^n q^{n^2}\right)^3 \sum_{m=-\infty}^{\infty} z^{-m} q^{\binom{m}{2}}}{(q)_\infty^3}$$

$$= \frac{(q^2; q^2)_\infty^3}{(q)_\infty^2} [z^0](zq; q^2)_\infty^3 (z^{-1}q; q^2)_\infty^3 (-z^{-1})_\infty (-zq)_\infty$$

(by [3, (2.2.10)])

$$= \frac{(q^2; q^2)_\infty^3}{(q)_\infty^2} [z^0](zq; -q)_\infty(-z^{-1}; -q)_\infty(zq; q^2)_\infty(z^{-1}q; q^2)_\infty$$

$$\times (z^2 q^2; q^4)_\infty(z^{-2}q^2; q^4)_\infty$$

$$= \frac{(q^2; q^2)_\infty^3}{(q)_\infty^2} [z^0] \frac{\sum_{n=-\infty}^\infty z^{-n}(-q)^{\binom{n}{2}} \sum_{m=-\infty}^\infty (-z)^m q^{m^2}}{\sum_{j=-\infty}^\infty (-1)^j z^{2j} q^{2j^2}} {(-q; -q)_\infty(q^2; q^2)_\infty(q^4; q^4)_\infty}$$

(by [3, (2.2.10)])

$$= \frac{1}{(q)_\infty(q^2; q^2)_\infty} \sum_{n, j=-\infty}^\infty (-q)^{\binom{n}{2}}(-1)^{n-2j} q^{(n-2j)^2+2j^2}(-1)^j .$$

Now replacing q by $-q$ in the above and simplifying, we arrive at the desired identity (3.2). □

Theorem 5. With $S_3(q)$ as defined in (3.1),

$$S_3(q) = T_1(q) - q^4 T_2(q) + 2q^3 T_3(q),$$

where

$$T_1(q) = \frac{(q^{28}, q^{32}, q^{60}; q^{60})_\infty}{(q^2; q^2)_\infty^2(q^3; q^3)_\infty(q^3; q^6)_\infty(q^2; q^4)_\infty},$$

$$T_2(q) = \frac{(q^8, q^{52}, q^{60}; q^{60})_\infty}{(q^2; q^2)_\infty^2(q^3; q^3)_\infty(q^3; q^6)_\infty(q^2; q^4)_\infty},$$

and

$$T_3(q) = \frac{(q^3; q^3)_\infty(q^{12}, q^{48}, q^{60}; q^{60})_\infty}{(q)_\infty(q^2; q^2)_\infty(q^6; q^{12})_\infty}.$$

Proof. Moving the infinite products to the left side of (3.2) and replacing j by $3j+v$ ($v = -1, 0, 1$), we see that

$$S_3(-q)(-q; -q)_\infty(q^2; q^2)_\infty$$

$$= \sum_{j,n=-\infty}^\infty \sum_{v=-1}^1 (-1)^{j+v} q^{n(3n-1)/2-4n(3j+v)+6(3j+v)^2}$$

$$= \sum_{j,n=-\infty}^\infty \sum_{v=-1}^1 (-1)^{j+v} q^{6v^2-4nv+20jv+n(3n-1)/2+30j^2-2j}$$

(where we have shifted n to $n + 4j$)

$$= - \sum_{j,n=-\infty}^{\infty} (-1)^j q^{n(3n+7)/2+30j^2-22j+6}$$

$$+ \sum_{j,n=-\infty}^{\infty} (-1)^j q^{n(3n-1)/2+30j^2-2j}$$

$$- \sum_{j,n=-\infty}^{\infty} (-1)^j q^{n(3n-9)/2+30j^2+18j+6}$$

$$= - q^4 \sum_{j,n=-\infty}^{\infty} (-1)^j q^{n(3n+1)/2+30j^2-22j}$$

$$+ \sum_{j,n=-\infty}^{\infty} (-1)^j q^{n(3n-1)/2+30j^2-2j}$$

$$- q^3 \sum_{j,n=-\infty}^{\infty} (-1)^j q^{3n(n-1)/2+30j^2+18j}$$

$$= \frac{(q^3;q^3)_\infty(-q)_\infty}{(-q^3;q^3)_\infty} ((q^{28},q^{32},q^{60};q^{60})_\infty - q^4(q^8,q^{52},q^{60};q^{60})_\infty)$$

$$- \frac{2q^3(q^6;q^6)_\infty}{(q^3;q^6)_\infty} (q^{12},q^{48},q^{60};q^{60})_\infty \quad \text{(by [3, (2.2.10)]).}$$

Now if we replace q by $-q$ in the above and isolate $S_3(q)$, we find that we have the result stated in Theorem 4. □

A. Berkovich [11] has pointed out that the triple theta series can be reduced to a linear combination of four infinite products utilizing his new proof of the tripentagonal number theorem and four identities from L. J. Slater's compendium [27, (19), (20), (34), (98)].

It should also be noted that Berkovich's proof [11] of the tripentagonal number theorem begins with a marvelous reduction of the triple q-series to a double q-series and thence to the Slater identities just listed.

4 The Triple q-Series

The genesis of this paper lies in the desire to understand the left-hand side of (1.2) in some partition-theoretic sense. We define

$$W_3(n) = \sum_{i,j,k \geq 0} \frac{q^{i^2+j^2+k^2}}{(q)_{i-j+k}(q)_{j-k+i}(q)_{k-i+j}}. \tag{4.1}$$

To this end, we change the indices of summation to m, n, r with

$$m = i - j + k,$$
$$n = j - k + i,$$
$$r = k - i + j,$$

and we deduce directly that

$$i = \frac{m + n}{2},$$
$$j = \frac{n + r}{2},$$
$$k = \frac{r + m}{2}.$$

This one-to-one transformation is admissible in triples of positive integers if and only if m, n, and r are all of the same parity.

Additionally, under this transformation,

$$i^2 + j^2 + k^2 = m \left(\frac{m + n}{2} \right) + n \left(\frac{n + r}{2} \right) + r \left(\frac{r + n}{2} \right).$$

Hence

$$W_3(n) = {\sum_{m,n,r \geq 0}}' \frac{q^{m(\frac{m+n}{2}) + n(\frac{n+r}{2}) + r(\frac{r+n}{2})}}{(q)_m (q)_n (q)_r}, \qquad (4.2)$$

where $\sum'$ means that the indices m, n, r must all have the same parity.

We now recall a basic fact about partition-generating functions (cf. [3, Chapter 1], [22, Chapter XIX]):

$$\frac{q^{An}}{(q)_n} \quad (A > 0)$$

is the generating function for partitions into exactly n parts each $\geq A$.

The above observations now allow us to interpret (4.1) in terms of tripartitions. We say that a multipartition $(\lambda_1, \lambda_2, \ldots, \lambda_k)$ has *equiparity* if each of $(\lambda_1, \lambda_2, \ldots, \lambda_k)$ has an even number of parts or each has an odd number. In a multipartition $(\lambda_1, \lambda_2, \ldots, \lambda_k)$ we say that λ_{i+1} is the *next* component to λ_i with the convention that λ_1 is next to λ_k (i.e., $\lambda_{k+1} = \lambda_1$). We say that a multipartition of equiparity has parts of *average size* if each part in component λ_i is at least as large as the average of the number of parts in λ_i and λ_{i+1}.

Finally, we define $B_3(n)$ to be the number of tripartitions of equiparity and parts of average size.

Theorem 6.

$$\sum_{n=0}^{\infty} B_3(n) q^n = W_3(q) = \sum_{i,j,k \geq 0} \frac{q^{i^2 + j^2 + k^2}}{(q)_{i-j+k} (q)_{j-k+i} (q)_{k-i+j}}.$$

Proof. Inspection of (4.2) reveals that $W_3(q)$ is indeed the generating function in question. $\square$

The obvious relevance of tripartitions with equiparity and average size suggests that we apply these ideas to other classes of partitions. We apply them to bipartitions in the next section.

5 Bipartitions with Parts of Average Size

In treating bipartitions, it turns out that we must keep track of whether the parity implied by equiparity is even or odd.

We define $B_{2,e}(n)$ (respectively $B_{2,0}(n)$) to be the number of bipartitions of even (respectively odd) equiparity with parts of average size.

Theorem 7.
$$B_{2,e}(n) - B_{2,0}(n) = \begin{cases} 0 & \text{if } n \text{ is odd,} \\ A_{2,2}\left(\frac{n}{2}\right) & \text{if } n \text{ is even,} \end{cases}$$

where $A_{2,2}(n)$ is the Rogers–Ramanujan partition function, the number of ordinary partitions of n into parts $\equiv \pm 1 \pmod 5$.

Proof.

$$\sum_{n=0}^{\infty} (B_{2,e}(n) - B_{2,0}(n))q^n$$

$$= \frac{1}{2} \sum_{n,m \geq 0} \frac{q^{n\left(\frac{n+m}{2}\right)+m\left(\frac{n+m}{2}\right)}}{(q)_n (q)_m} ((-1)^n + (-1)^m)$$

$$= \sum_{n,m \geq 0} \frac{(-1)^n q^{n\left(\frac{n+m}{2}\right)+m\left(\frac{n+m}{2}\right)}}{(q)_n (q)_m}$$

$$= \sum_{n,m \geq 0} \frac{(-1)^n q^{\frac{n^2}{2}+\frac{m^2}{2}+mn}}{(q)_n (q)_m}$$

$$= \sum_{n \geq 0} \frac{q^{2n^2}}{(q^2; q^2)_n} \quad \text{(by [30, p. 45])}$$

$$= \prod_{n=1}^{\infty} \frac{1}{(1-q^{10n-2})(1-q^{10n-8})} \quad \text{(by [3, (7.1.7)])}$$

$$= \sum_{n=0}^{\infty} A_{2,2}(n)q^{2n}.$$

Comparison of the extremes in the above string of equations yields the assertion of our theorem. □

It is natural to ask why we examined the excess of even equiparity over odd equiparity. Why not just consider the full count of such partitions? This can be done of course, however, the resulting generating functions reduce to series such as

$$\sum_{j=0}^{\infty} \frac{(-1)^j q^{j^2}}{(q)_{2j}},$$

which appear not to have the interesting qualities associated with the Rogers–Ramanujan identities or related formulas.

6 Further Multipartition Identities

Once we have introduced the idea of multipartitions with average size, it is natural to consider seemingly simpler questions. For example, let us consider $C_k(n)$, the number of k-component multipartitions in which each part in the ith component $(1 \leq i \leq k)$ is larger than the number of parts in the next component.

Lemma 8.

$$\sum_{j \geq 0} (q^{j+1})_j q^j = \sum_{n=0}^{\infty} (-1)^n q^{n(15n+7)/2} + \sum_{n=0}^{\infty} (-1)^n q^{n(15n+13)/2 \ +1}$$

$$+ \sum_{n=0}^{\infty} (-1)^n q^{n(15n+17)/2 \ +2} + \sum_{n=0}^{\infty} (-1)^n q^{n(15n+23)/2 \ +4}.$$

Proof.

$$\sum_{j \geq 0} q^j (q^{j+1})_j$$

$$= \sum_{j \geq 0} q^j \sum_{h=0}^{j} \begin{bmatrix} j \\ h \end{bmatrix} (-1)^h q^{\binom{h+1}{2}+jh} \quad \text{(by [3, (3.3.6)])}$$

$$= \sum_{h,j \geq 0} \begin{bmatrix} j+h \\ h \end{bmatrix} (-1)^h q^{\binom{h+1}{2}+(j+h)h+j+h}$$

$$= \sum_{h \geq 0} \frac{(-1)^h q^{3h(h+1)/2}}{(q^{h+1})_{h+1}} \quad \text{(by [3, (3.3.7)])}$$

$$= \sum_{n=0}^{\infty}(-1)^n q^{n(15n+7)/2} + \sum_{n=0}^{\infty}(-1)^n q^{n(15n+13)/2} \ +1$$

$$+ \sum_{n=0}^{\infty}(-1)^n q^{n(15n+17)/2} \ +2 + \sum_{n=0}^{\infty}(-1)^n q^{n(15n+23)/2} \ +4$$

(by [26, A(8), p. 333]),

which is the assertion of the lemma. □

Theorem 9.

$$\sum_{n\geq 0} C_2(n)q^n = \frac{1}{(q)_\infty}\left(\sum_{n=0}^{\infty}(-1)^n q^{n(15n+7)/2}(1 + q^{3n+1} + q^{5n+2} + q^{8n+4})\right).$$

Proof. Clearly,

$$\sum_{n\geq 0} C_2(n)q^n = \sum_{i,j\geq 0} \frac{q^{i(j+1)+j(i+1)}}{(q)_i(q)_j}$$

$$= \sum_{j\geq 0} \frac{q^j}{(q)_j(q^{2j+1})_\infty} \quad \text{(by [3, (2.2.5)])}$$

$$= \frac{1}{(q)_\infty} \sum_{j\geq 0} q^j(q^{j+1})_j,$$

and we replace the final sum by the expression in Lemma 8, and the result follows.

□

For our next result, we recall one of Ramanujan's fifth-order mock theta functions [25, p. 131], [29, p. 278]:

$$\chi_1(q) = \sum_{i\geq 0} \frac{q^i}{(q^{i+1})_{i+1}}.$$

Theorem 10. $\sum_{n\geq 0} C_3(n)q^n = \frac{\chi_1(q)}{(q)_\infty}.$

Proof. Clearly

$$\sum_{n\geq 0} C_3(n)q^n = \sum_{i,j,k\geq 0} \frac{q^{i(j+1)+j(k+1)+k(i+1)}}{(q)_i(q)_j(q)_k}$$

$$= \sum_{i,j\geq 0} \frac{q^{ij+i+j}}{(q)_i(q)_j(q^{j+i+1})_\infty} \quad \text{(by [3, (2.2.5)])}$$

$$= \frac{1}{(q)_\infty} \sum_{i,j \geq 0} \begin{bmatrix} j+i \\ i \end{bmatrix} q^{ij+i+j}$$

$$= \frac{1}{(q)_\infty} \sum_{i \geq 0} \frac{q^i}{(q^{i+1})_{i+1}} \qquad \text{(by [3, (3.3.7)])}$$

$$= \frac{\chi_1(q)}{(q)_\infty}.$$

$\square$

Theorem 11. $\sum_{n \geq 0} C_4(n)q^n = \frac{1}{(q)_\infty} \sum_{i,j \geq 1} \frac{q^{ij-1}}{(q)_{i+j-1}}$.

Proof. As previously,

$$\sum_{n \geq 0} C_4(n)q^n = \sum_{i,j,k,m \geq 0} \frac{q^{i(j+1)+j(k+1)+k(m+1)+m(i+1)}}{(q)_i (q)_j (q)_k (q)_m}$$

$$= \sum_{i,j,k \geq 0} \frac{q^{ij+jk+i+j+k}}{(q)_i (q)_j (q)_k (q^{k+i+1})_\infty} \qquad \text{(by [3, (2.2.5)])}$$

$$= \frac{1}{(q)_\infty} \sum_{i,j,k \geq 0} \begin{bmatrix} k+i \\ i \end{bmatrix} \frac{1}{(q)_j} q^{ij+jk+i+j+k}$$

$$= \frac{1}{(q)_\infty} \sum_{i,k \geq 0} \begin{bmatrix} k+i \\ i \end{bmatrix} \frac{q^{i+k}}{(q^{i+k+1})_\infty}$$

$$= \frac{1}{(q)_\infty^2} \sum_{i \geq 0} q^i (q)_i \sum_{k \geq 0} \frac{(q^{i+1})_k^2 q^k}{(q)_k}$$

$$= \frac{1}{(q)_\infty^2} \sum_{i \geq 0} q^i (q)_i \frac{(q^{i+1})_\infty (q^{i+2})_\infty}{(q)_\infty} \sum_{j \geq 0} \frac{q^{(i+1)j}}{(q^{i+2})_j}$$

$$= \frac{1}{(q)_\infty} \sum_{i,j \geq 0} \frac{q^{ij+i+j}}{(q)_{j+i+1}}$$

$$= \frac{1}{(q)_\infty} \sum_{i,j \geq 1} \frac{q^{ij-1}}{(q)_{i+j-1}}.$$

$\square$

Theorem 12.

$$\sum_{n \geq 0} C_5(n)q^n = \frac{1}{(q)_\infty^2} \sum_{m=1}^\infty \sigma(m)q^{m-1} = e^{-13\pi i\tau/6}\eta^{-2}(\tau)(1 - E_2(\tau))/24,$$

where $\sigma(m)$ is the sum of the divisors of m, $q = e^{2\pi i\tau}$, $\eta(\tau)$ is Dedekind's η-function given by

$$\eta(\tau) = q^{1/24}(q)_\infty,$$

and $E_2(\tau)$ is the Eisenstein series given by

$$E_2(\tau) = 1 - 24 \sum_{m=1}^\infty \sigma(m)q^m.$$

Proof.

$$\sum_{n \geq 0} C_5(n)q^n$$

$$= \sum_{i,j,k,\ell,m \geq 0} \frac{q^{ij+jk+k\ell+\ell m+mi+i+j+k+\ell+m}}{(q)_i(q)_j(q)_k(q)_\ell(q)_m}$$

$$= \sum_{i,k,m \geq 0} \frac{q^{mi+i+k+m}}{(q)_m(q)_k(q)_i(q^{i+k+1})_\infty(q^{k+m+1})_\infty} \quad \text{(by [3, (2.2.5)])}$$

$$= \frac{1}{(q)_\infty} \sum_{k,m \geq 0} \frac{q^{k+m}}{(q)_m(q^{k+m+1})_\infty} \sum_{i \geq 0} \frac{(q^{k+1})_i q^{i(m+1)}}{(q)_i}$$

$$= \frac{1}{(q)_\infty} \sum_{k,m \geq 0} \frac{q^{k+m}(q^{m+k+2})_\infty}{(q)_m(q^{k+m+1})_\infty(q^{m+1})_\infty} \quad \text{(by [3, (2.2.1)])}$$

$$= \frac{1}{(q)_\infty^2} \sum_{k,m \geq 0} \frac{q^{k+m}}{(1 - q^{k+m+1})}$$

$$= \frac{1}{(q)_\infty^2} \sum_{N=0}^\infty \frac{q^N(N+1)}{(1 - q^{N+1})}$$

$$= \frac{q^{-1}}{(q)_\infty^2} \sum_{m=1}^\infty \frac{mq^m}{1 - q^m}$$

$$= \frac{q^{-1}}{(q)_\infty^2} \sum_{m=1}^\infty \sigma(m)q^m.$$

$\square$

Corollary 13.

$$\sum_{m=1}^{\infty} \sigma(m)q^m \equiv q(q)_{\infty}^2 H(q^5) \pmod 5,$$

where

$$H(q) = 1/((q^2; q^5)_{\infty}(q^3; q^5)_{\infty}).$$

Proof. In examining the quintuple series-generating function for $C_5(n)$, we see that cyclic permutation of the five indices leaves each term unaltered. Since there are no nontrivial subgroups of the cyclic group of order 5, we see that modulo 5,

$$\sum_{n=0}^{\infty} C_5(n)q^n \equiv \sum_{n=0}^{\infty} \frac{q^{5n^2+5n}}{(q)_n^5} \pmod 5$$

$$\equiv \sum_{n=0}^{\infty} \frac{q^{5n^2+5n}}{(q^5; q^5)_n} = H(q^5) \quad \text{(by [3, (7.1.7)])}.$$

The result now follows from Theorem 12.   □

7 Another Family of Mulitpartition Identities

The partition function $C_k(n)$ is the number of multipartitions of n in which each part in the ith component is larger than the number of parts in the next component. In this section we consider $D_k(n)$, the number of multipartitions of n with k components in which each part in the ith component is larger that the number of parts in the next component with the exception of the kth component (which now does *not* have to have its parts larger than the number of elements in the first component).

Theorem 14. $\sum_{n\geq0} D_2(n)q^n = \frac{1}{(1-q)(q)_{\infty}}.$

Proof. $\sum_{n\geq0} D_2(n)q^n = \sum_{i,j\geq0} \frac{q^{i(j+1)+j}}{(q)_i(q)_j} = \frac{1}{(q)_{\infty}} \sum_{i\geq0} q^i = \frac{1}{(1-q)(q)_{\infty}}.$ □

Theorem 15. $\sum_{n\geq0} D_3(n)q^n = \frac{q^{-1}}{(q)_{\infty}^2} - \frac{q^{-1}}{(q)_{\infty}}.$

Proof.

$$\sum_{n\geq0} D_3(n)q^n$$

$$= \sum_{i,j,k\geq0} \frac{q^{i(j+1)+j(k+1)+k}}{(q)_i(q)_j(q)_k}$$

$$= \frac{1}{(q)_{\infty}} \sum_{i,j\geq0} \frac{q^{i(j+1)+j}}{(q)_i}$$

$$= \frac{1}{(q)_\infty} \sum_{i \geq 0} \frac{q^i}{(q)_{i+1}}$$

$$= \frac{q^{-1}}{(q)_\infty} \sum_{i \geq 1} \frac{q^i}{(q)_i}$$

$$= \frac{q^{-1}}{(q)_\infty} \left(\frac{1}{(q)_\infty} - 1 \right).$$

$\square$

Theorem 16. $\sum_{n \geq 0} D_4(n) q^n = \frac{q^{-1} \Delta(q)}{(q)_\infty^2}$, where $\Delta(q) = \sum_{n=1}^\infty \frac{q^n}{(1-q^n)}$ is the generating function for the number of divisors of n.

Proof.

$$\sum_{n \geq 0} D_4(n) q^n$$

$$= \sum_{i,j,k,\ell \geq 0} \frac{q^{i(j+1)+j(k+1)+k(\ell+1)+\ell}}{(q)_i (q)_j (q)_k (q)_\ell}$$

$$= \frac{1}{(q)_\infty} \sum_{i,j,k \geq 0} \frac{q^{i(j+1)+j(k+1)+k}}{(q)_i (q)_j}$$

$$= \frac{1}{(q)_\infty} \sum_{i,j} \frac{q^{i(j+1)+j}}{(q)_i (q)_{j+1}}$$

$$= \frac{1}{(q)_\infty^2} \sum_{j \geq 0} \frac{q^j}{(1 - q^{j+1})}$$

$$= \frac{q^{-1}}{(q)_\infty^2} \Delta(q).$$

$\square$

Theorem 17. $\sum_{n \geq 0} D_5(n) q^n = \frac{1}{(q)_\infty^2} \sum_{k \geq 1} \frac{q^{k-1}}{(q)_k (1-q^k)}.$

Proof.

$$\sum_{n \geq 0} D_5(n) q^n$$

$$= \sum_{i,j,k,\ell,m \geq 0} \frac{q^{i(j+1)+j(k+1)+k(\ell+1)+\ell(m+1)+m}}{(q)_i (q)_j (q)_k (q)_\ell (q)_m}$$

$$= \frac{1}{(q)_\infty} \sum_{i,j,k \geq 0} \frac{q^{ij+jk+i+j+k}}{(q)_i (q)_j (q)_{k+1}}$$

$$= \frac{1}{(q)_\infty^2} \sum_{j,k \geq 0} \frac{q^{jk+j+k}}{(q)_{k+1}}$$

$$= \frac{1}{(q)_\infty^2} \sum_{k \geq 0} \frac{q^k}{(q)_{k+1}(1 - q^{k+1})} .$$

□

Theorem 18. $\sum_{n \geq 0} D_6(n) q^n = \frac{q^{-1}}{(q)_\infty^3} \sum_{h \geq 0} \frac{1 - (q^{h+1})_\infty}{1 - q^{h+1}}$.

Proof.

$$\sum_{n \geq 0} D_6(n) q^n$$

$$= \sum_{i,j,k,h,m,n \geq 0} \frac{q^{i(j+1)+j(h+1)+k(n+1)+h(m+1)+m(n+1)+n}}{(q)_i (q)_j (q)_k (q)_h (q)_m (q)_n}$$

$$= \frac{1}{(q)_\infty^2} \sum_{j,k,h,m \geq 0} \frac{q^{jk+kh+hm+j+k+h+m}}{(q)_k (q)_h}$$

$$= \frac{q^{-1}}{(q)_\infty^2} \sum_{k,h \geq 0} \frac{q^{(k+1)(h+1)}}{(q)_{k+1}(q)_{h+1}}$$

$$= \frac{q^{-1}}{(q)_\infty^2} \sum_{k,h \geq 1} \frac{q^{kh}}{(q)_k (q)_h}$$

$$= \frac{q^{-1}}{(q)_\infty^2} \sum_{h \geq 1} \frac{1}{(q)_h} \left(\frac{1}{(q^h)_\infty} - 1 \right)$$

$$= \frac{q^{-1}}{(q)_\infty^3} \sum_{h \geq 1} \frac{1 - (q^h)_\infty}{1 - q^h} .$$

□

8 Interpretations of Multiple q-Series Identities

There are numerous theorems identifying multiple q-series including [1], [2], [4], [5], [24]. Whether this approach using multipartitions will provide really new insight concerning these identities awaits future investigations. In this section, we choose one example from [4] that has not been given a partition-theoretic interpretation previously.

We define $W(n)$ to be the number of tripartitions $\lambda_1 + \lambda_2 + \lambda_3$ of n for which the following three conditions are met (here #(λ_i) is the number of parts in λ_i):

   (i) each part in λ_1 is $\geq$ #(λ_1);

   (ii) each part in λ_2 is $\geq$ #(λ_1) + #(λ_2);

   (iii) each part in λ_3 is ≥ 2(#(λ_1) + #(λ_2) + #(λ_3)).

We define $V(n)$ to be the number of bipartitions $\lambda_1 + \lambda_2$ wherein all parts of both components are $\equiv \pm 1 \pmod 5$.

Theorem 19. *For each $n \geq 0$, $W(n) = V(n)$.*

Proof. In [4, (1.7)], it is shown that

$$\sum_{n,m,r \geq 0} \frac{q^{(n+m)^2+(r+m)^2+nr}}{(q)_n(q)_m(q)_r} = \frac{1}{(q,q^4;q^5)_\infty^2}.$$

Clearly [3, (7.1.6)],

$$\sum_{n=0}^{\infty} V(n)q^n = \frac{1}{(q,q^4;q^5)_\infty^2}$$

and

$$\sum_{n,m,r \geq 0} \frac{q^{(n+m)^2+(r+m)^2+nr}}{(q)_n(q)_m(q)_r} = \sum_{n,m,r \geq 0} \frac{q^{n^2+r(r+n)+m(2m+2r+2n)}}{(q)_n(q)_m(q)_r} = \sum_{n \geq 0} W(n)q^n,$$

and our result follows. $\qquad\qquad\qquad\qquad\qquad\qquad\qquad\qquad\qquad\qquad$ $\square$

References

1. G. E. Andrews, An analytic generalization of the Rogers-Ramanujan identities for odd moduli, Proc. Nat. Acad. Sci. USA, 71, (1974), 4082–4085.
2. G. E. Andrews, *Problems and prospects for basic hypergeometric functions*, from The Theory and Applications of Special Functions (R. Askey, ed.), pp. 191–224, Academic Press, New York, 1975.
3. G. E. Andrews, The Theory of Partitions, Encycl. of Math. and Its Appl. (G.-C. Rota, ed.), Vol. 2, Addison-Wesley, Reading, 1975 (Reprinted: Cambridge University Press, Cambridge, 1998).
4. G. E. Andrews, Multiple q-series identities, Houston Math. J., 7 (1981), 11–22.
5. G. E. Andrews, Multiple series Rogers-Ramanujan identities, Pac. J. Math., 114 (1984), 267–283.

6. G. E. Andrews, Umbral calculus, Bailey chains and pentagonal number theorems, J. Comb. Th., Ser. A, 91 (2000), 464–475.
7. G. E. Andrews and F. G. Garvan, Dyson's crank of a partition, Bull. Amer. Math. Soc., 18 (1988), 167–171.
8. G. E. Andrews and R. Roy, Ramanujan's method in q-series congruences, Elec. J. Comb., 4(2): R2, 7pp., 1997.
9. A. O. L. Atkin, Proof of a conjecture of Ramanujan, Glasgow Math. J., 8 (1967), 14–32.
10. A. O. L. Atkin, Ramanujan congruences for $p_{-k}(n)$, Canad. J. Math., 20 (1968), 67–78.
11. A. Berkovich, The tripentagonal number theorem and related identities, Int. J. Number Theory, (to appear).
12. P. Bouwknegt, Multipartitions, generalized Durfee squares and affine Lie algebra characters, J. Austral. Math. Soc., 72 (2002), 395–408.
13. M. Broué and G. Malle, Zyklotomische Heckealgebren, Astérisque, 212 (1993), 119–189.
14. M. S. Cheema and C. T. Haskell, Multirestricted and rowed partitions, Duke Math. J., 34 (1967), 443–451.
15. F. J. Dyson, Some guesses in the theory of partitions, Eureka (Cambridge), 8 (1944), 10–15.
16. M. Fayers, Weights of multipartitions and representations of Ariki-Koike algebras, Adv. in Math., 206 (2006), 112–144.
17. F. G. Garvan, New combinatorial interpretations of Ramanujan's partition congruences mod 5, 7 and 11, Trans. Amer. Math. Soc., 305 (1988), 47–77.
18. F. G. Garvan, Ranks and cranks for bipartitions mod 5, (in preparation).
19. G. Gasper and N. Rahman, *Basic Hypergeometric Series*, Cambridge University Press, Cambridge, 1990.
20. H. Gupta, *Selected Topics in Number Theory*, Abacus Press, Turnbridge Wells, 1980.
21. H. Gupta, C. E. Gwyther, and J. C. P. Miller, *Tables of Partitions*, Royal Society Math. Tables, Vol. 4, Cambridge University Press, Cambridge, 1958.
22. G. H. Hardy and E. M. Wright, *An Introduction to the Theory of Numbers*, 4th ed., Oxford University Press, London, 1960.
23. K. Mahlburg, Partition congruences and the Andrews-Garvan-Dyson crank, Proc. Nat. Acad. Sci., U.S.A, 102 (2005), 15373–15376.
24. P. Paule, On identities of the Rogers–Ramanujan type, J. Math. Anal. and Appl., 107 (1985), 255–284.
25. S. Ramanujan, *The Lost Notebook and Other Unpublished Papers*, intro. by G. E. Andrews, Narosa, New Delhi, 1987.
26. L. J. Rogers, On two theorems of combinatory analysis and allied identities, Proc. London Math. Soc. (2), 16 (1917), 315–336.
27. L. J. Slater, Further identities of the Rogers-Ramanujan type, Proc. London Math. Soc. (2), 54 (1952), 147–167.
28. J. J. Sylvester, A constructive theory of partitions arranged in three acts, an interact and an exodion, Amer. J. Math., 5 (1882), 251–330.
29. G. N. Watson, The mock theta functions (2), Proc. London Math. Soc., Ser. 2, 42 (1937), 274–304.
30. G. N. Watson, A note on Lerch's functions, Quart. J. Math., Oxford Series, 8 (1937), 44–47.

2

Koshliakov's Formula and Guinand's Formula in Ramanujan's Lost Notebook

Bruce C. Berndt[1,*], Yoonbok Lee[2], and Jaebum Sohn[3,**]

[1]Department of Mathematics, University of Illinois, 1409 West Green Street,
Urbana, IL 61801, USA
berndt@math.uiuc.edu
[2]Department of Mathematics, Yonsei University, 134 Shinchon-dong, Seodaemun-gu,
Seoul, 120-749, Korea
leeyb@yonsei.ac.kr
[3]Department of Mathematics, Yonsei University, 134 Shinchon-dong, Seodaemun-gu,
Seoul, 120-749, Korea
jsohn@yonsei.ac.kr

Summary. On two pages in his lost notebook, Ramanujan recorded several theorems involving the modified Bessel function $K_\nu(z)$. These include Koshliakov's formula and Guinand's formula, both connected with the functional equation of nonanalytic Eisenstein series, and both discovered by these authors several years after Ramanujan's death. Other formulas, including one by K. Soni and two particularly elegant new results, are also stated without proof by Ramanujan. In this paper, we prove all the formulas claimed by Ramanujan on these two pages and conclude with a survey of related results.

Key words: Koshliakov's formula, Guinand's formula, modified Bessel functions, nonanalytic Eisenstein series, Maass wave forms, divisor functions.

2000 *Mathematics Subject Classifications*: Primary 33C10; Secondary 11F12, 11F20

1 Introduction

At a conference held on June 1–5, 1987, at the University of Illinois to commemorate the centenary of Ramanujan's birth, R. William Gosper remarked in his lecture that Ramanujan frequently reaches his hand from his grave to snatch your theorems from you. In less colorful language, Gosper asserted that it frequently happens that one proves an important theorem, which later is found to be ensconced somewhere in Ramanujan's writings. In other instances, we learn that Ramanujan had anticipated an important recent development in his own inimitable way.

*Research partly supported by NSA grant MSPF-03IG-124.
**Research supported by Korea Research Foundation Grant KRF-2003-015-C00008.

K. Alladi (ed.), *Surveys in Number Theory*, DOI: 10.1007/978-0-387-78510-3_2,
© Springer Science+Business Media, LLC 2008

In this paper, we examine two pages in Ramanujan's lost notebook [28, pp. 253–254] on which Gosper's observation is demonstrated once again. On page 253, Ramanujan states a version of A.P. Guinand's formula, from which N.S. Koshliakov's formula follows as a corollary. Although there is no indication that Ramanujan made any connections with Epstein zeta functions or nonholomorphic Eisenstein series, it is remarkable that with these two important formulas, Ramanujan anticipated later developments that are now central in the theory of Maass wave forms on SL(2, $\mathbb{Z}$). On page 254, Ramanujan gives applications of Guinand's formula; these results are mostly new.

Koshliakov is chiefly remembered for one theorem, namely, Koshliakov's formula [21], which we now see was proved by Ramanujan about ten years earlier. To state this formula, let $K_\nu(z)$ denote the modified Bessel function of order ν [37, p. 78], and let $d(n)$ denote the number of positive divisors of the positive integer n. Then, if γ denotes Euler's constant and $a > 0$,

$$\gamma - \log\left(\frac{4\pi}{a}\right) + 4\sum_{n=1}^{\infty} d(n)K_0(2\pi a n)$$

$$= \frac{1}{a}\left(\gamma - \log(4\pi a) + 4\sum_{n=1}^{\infty} d(n)K_0\left(\frac{2\pi n}{a}\right)\right). \tag{1.1}$$

Koshliakov's proof, as well as most subsequent proofs, depends upon Voronoï's summation formula [35]

$$\sum_{a \le n \le b}{}' d(n)f(n) = \int_a^b (\log x + 2\gamma)f(x)dx$$

$$+ \sum_{n=1}^{\infty} d(n)\int_a^b f(x)(4K_0(4\pi\sqrt{nx}) - 2\pi Y_0(4\pi\sqrt{nx}))dx,$$

$$\tag{1.2}$$

where $Y_\nu(z)$ denotes the Weber–Bessel function of order ν, usually so denoted [37, p. 64]. The prime $'$ on the summation sign on the left-hand side indicates that if a or b is an integer, then only $\frac{1}{2}f(a)$ or $\frac{1}{2}f(b)$, respectively, is counted. For conditions on $f(x)$ that ensure the validity of (1.2), see, for example, Berndt's paper [4].

A.L. Dixon and W.L. Ferrar [12] also proved (1.1) using the Voronoï summation formula. F. Oberhettinger and K.L. Soni [26] established a generalization of (1.1) using Voronoï's formula (1.2). Soni [32] derived further identities from Koshliakov's formula. In contrast to the work of these authors, Ramanujan evidently did not appeal to Voronoï's formula.

Koshliakov's formula can be considered as an analogue of the transformation formula for the classical theta function, namely,

$$\sum_{n=-\infty}^{\infty} e^{-\pi n^2/\tau} = \sqrt{\tau} \sum_{n=-\infty}^{\infty} e^{-\pi n^2 \tau}, \qquad \mathrm{Re}\,\tau > 0, \tag{1.3}$$

which, as is well known, is equivalent to the functional equation of the Riemann zeta function $\zeta(s)$ given by [34, p. 22]:

$$\pi^{-s/2}\Gamma\left(\frac{1}{2}s\right)\zeta(s) = \pi^{-(1-s)/2}\Gamma\left(\frac{1}{2}(1-s)\right)\zeta(1-s). \tag{1.4}$$

Ferrar [13] was evidently the first mathematician to prove that (1.1) can indeed be derived from the functional equation of $\zeta^2(s)$. Oberhettinger and Soni [26] showed that (1.4) and Koshliakov's formula are equivalent.

On page 253 in his lost notebook [28], Ramanujan states (1.1) as a corollary of a more general and especially beautiful formula at the top of the same page. This more general formula is stated in an equivalent formulation in Theorem 1.1 below.

Theorem 1.1 (p. 253). *Let $\sigma_k(n) = \sum_{d|n} d^k$, and let $\zeta(s)$ denote the Riemann zeta function. If α and β are positive numbers such that $\alpha\beta = \pi^2$, and if s is any complex number, then*

$$\sqrt{\alpha}\sum_{n=1}^{\infty}\sigma_{-s}(n)n^{s/2}K_{s/2}(2n\alpha) - \sqrt{\beta}\sum_{n=1}^{\infty}\sigma_{-s}(n)n^{s/2}K_{s/2}(2n\beta)$$

$$= \frac{1}{4}\Gamma\left(\frac{s}{2}\right)\zeta(s)\{\beta^{(1-s)/2} - \alpha^{(1-s)/2}\}$$

$$+ \frac{1}{4}\Gamma\left(-\frac{s}{2}\right)\zeta(-s)\{\beta^{(1+s)/2} - \alpha^{(1+s)/2}\}. \tag{1.5}$$

The identity (1.5) is equivalent to a formula established by Guinand [15] in 1955. The series in Theorem 1.1 are reminiscent of the Fourier expansion of nonanalytic Eisenstein series on $SL(2, \mathbb{Z})$, or Maass wave forms [23], [25, pp. 230–232], [22, pp. 15–16], [33, pp. 208–209]. This Fourier series was published by H. Maass [23] in the language of Eisenstein series in the same year, 1949, that A. Selberg and S. Chowla [30], [29, pp. 367–378] published it in the similar vein of the Epstein zeta function, but with their proof not published until several years later [31], [29, pp. 521–545]. In the meanwhile, P.T. Bateman and E. Grosswald [2] published a proof. These Eisenstein series were shown by Maass [23] to satisfy a functional equation for automorphic forms. C.J. Moreno kindly informed the authors that he was easily able to derive Theorem 1.1 from the aforementioned Fourier series expansion and functional equation. One may then regard (1.5) as an equivalent formulation of the functional equation of these nonholomorphic Eisenstein series or these particular Maass wave forms. S.P. Zwegers [38] has recently found connections between nonholomorphic modular forms and Ramanujan's mock theta functions. We have been unable to find evidence in Ramanujan's work that he made connections of (1.5) with nonholomorphic Eisenstein series or any other nonholomorphic modular forms. The proof of Theorem 1.1 that we give below is essentially the same as that of Guinand [15] and is completely independent of any considerations with nonanalytic Eisenstein series or their closely associated Epstein zeta functions. As is well known, Ramanujan made a large number of original contributions to *analytic* Eisenstein series, many of which can be found in his lost notebook [1], [9].

On page 254, Ramanujan records formulas similar to Koshliakov's formula (1.1) or to Guinand's formula (1.5). We show that each of the three main results on this page can be deduced from Ramanujan's (and Guinand's) beautiful generalization (1.5) of Koshliakov's formula.

We close this Introduction by mentioning two recent papers by S. Kanemitsu, Y. Tanigawa, H. Tsukada, and M. Yoshimoto [18] and S. Kanemitsu, Y. Tanigawa, and M. Yoshimoto [19], in which the formulas of Koshliakov and Guinand are used or generalized.

2 Preliminary Results

Throughout pages 253 and 254 of [28], Ramanujan expresses his theorems in terms of variants of the integral [14, p. 384, formula 3.471, no. 9]

$$\int_0^\infty x^{\nu-1} e^{-\beta/x - \gamma x} dx = 2 \left(\frac{\beta}{\gamma} \right)^{\nu/2} K_\nu(2\sqrt{\beta\gamma}), \tag{2.1}$$

where s is any complex number and Re $\beta > 0$, Re $\gamma > 0$. Since the modified Bessel function $K_\nu(z)$ is such a well-known function and its notation standard, it seems advisable to avoid Ramanujan's notation for variants of (2.1), which he calls ϕ, ψ, and χ. In summary, we have converted all of Ramanujan's theorems to identities involving the modified Bessel function K_ν.

We use the well-known fact [14, p. 978, formula 8.469, no. 3]

$$K_{1/2}(z) = \sqrt{\frac{\pi}{2z}} e^{-z}. \tag{2.2}$$

We require the simple asymptotic formula [37, p. 202]

$$K_\nu(z) \sim \sqrt{\frac{\pi}{2z}} e^{-z}, \quad z \to \infty,$$

to ensure the convergence of series and integrals and also to justify the interchange of integration and summation several times in the sequel. We need several integrals of Bessel functions beginning with [14, p. 705, formula 6.544, no. 8]

$$\int_0^\infty K_\nu\left(\frac{a}{x}\right) K_\nu(bx) \frac{dx}{x^2} = \frac{\pi}{a} K_{2\nu}(2\sqrt{ab}), \quad \text{Re } a > 0, \text{Re } b > 0. \tag{2.3}$$

We need the related pair [32, p. 544, equation (8)]

$$\int_0^\infty x K_0(ax) K_0(bx) dx = \frac{\log(a/b)}{a^2 - b^2}, \quad a, b > 0, \tag{2.4}$$

and [14, p. 697, formula 6.521, no. 3]

$$\int_0^\infty x K_\nu(ax) K_\nu(bx) dx = \frac{\pi (ab)^{-\nu}(a^{2\nu} - b^{2\nu})}{2\sin(\pi\nu)(a^2 - b^2)}, \quad |\mathrm{Re}\,\nu| < 1, \ \mathrm{Re}\,(a+b) > 0.$$

(2.5)

Lastly, we need the evaluation [14, p. 708, formula 6.561, no. 16], for Re $a > 0$ and Re$(\mu + 1 \pm \nu) > 0$,

$$\int_0^\infty x^\mu K_\nu(ax) dx = 2^{\mu-1} a^{-\mu-1} \Gamma\left(\frac{1+\mu+\nu}{2}\right) \Gamma\left(\frac{1+\mu-\nu}{2}\right). \quad (2.6)$$

3 Guinand's Formula

We begin by restating Theorem 1.1.

Entry 3.1 (p. 253). *As usual, let* $\sigma_k(n) = \sum_{d|n} d^k$, *and let* $\zeta(s)$ *denote the Riemann zeta function. If* α *and* β *are positive numbers such that* $\alpha\beta = \pi^2$, *and if* s *is any complex number, then*

$$\sqrt{\alpha} \sum_{n=1}^\infty \sigma_{-s}(n) n^{s/2} K_{s/2}(2n\alpha) - \sqrt{\beta} \sum_{n=1}^\infty \sigma_{-s}(n) n^{s/2} K_{s/2}(2n\beta)$$

$$= \frac{1}{4} \Gamma\left(\frac{s}{2}\right) \zeta(s) \{\beta^{(1-s)/2} - \alpha^{(1-s)/2}\}$$

$$+ \frac{1}{4} \Gamma\left(-\frac{s}{2}\right) \zeta(-s) \{\beta^{(1+s)/2} - \alpha^{(1+s)/2}\}. \quad (3.1)$$

To prove Entry 3.1, we need the following lemma.

Lemma 3.2. *Let* $K_\nu(z)$ *denote the modified Bessel function of order* ν. *If* $x > 0$ *and* Re $\nu > 0$, *then*

$$\frac{1}{4} (\pi x)^{-\nu} \Gamma(\nu) + \sum_{n=1}^\infty n^\nu K_\nu(2\pi n x)$$

$$= \frac{1}{4} \sqrt{\pi} (\pi x)^{-\nu-1} \Gamma\left(\nu + \frac{1}{2}\right) + \frac{\sqrt{\pi}}{2x} \left(\frac{x}{\pi}\right)^{\nu+1} \Gamma\left(\nu + \frac{1}{2}\right) \sum_{n=1}^\infty (n^2 + x^2)^{-\nu-1/2}.$$

(3.2)

Lemma 3.2 is due to G.N. Watson [36], who proved it by using the Poisson summation formula. H. Kober [20] generalized Lemma 3.2 in two different directions. In one of them, the index n on the left-hand side of (3.2) was replaced by $n + \alpha, 0 < \alpha < 1$, and in the other, $\cos(2\pi n\beta)$ was introduced into the summands on the left-hand side of (3.2). Berndt [5] generalized (3.2) by putting either even or odd periodic coefficients in the infinite series of (3.2). The proof that we give below is essentially an elaboration of Guinand's proof [15].

Proof of Entry 3.1. Setting $u = x^2$, using (2.1), and setting $n = kd$, we find that

$$\sqrt{\alpha} \sum_{n=1}^{\infty} \sigma_{-s}(n) n^{s/2} K_{s/2}(2n\alpha) = \sqrt{\alpha} \sum_{n=1}^{\infty} \sum_{d|n} d^{-s} n^{s/2} K_{s/2}(2n\alpha)$$

$$= \sqrt{\alpha} \sum_{d=1}^{\infty} \sum_{k=1}^{\infty} \left(\frac{k}{d}\right)^{s/2} K_{s/2}(2dk\alpha). \qquad (3.3)$$

We now invoke Lemma 3.2 on the right-hand side above to deduce that for Re $s > 0$,

$$\sqrt{\alpha} \sum_{n=1}^{\infty} \sigma_{-s}(n) n^{s/2} K_{s/2}(2n\alpha)$$

$$= \sqrt{\alpha} \sum_{d=1}^{\infty} \frac{1}{d^{s/2}} \left(-\frac{1}{4}(d\alpha)^{-s/2} \Gamma\left(\frac{s}{2}\right) + \frac{1}{4}\sqrt{\pi}(d\alpha)^{-s/2-1}\Gamma\left(\frac{s+1}{2}\right)\right.$$

$$\left. + \frac{\pi^{3/2}}{2d\alpha} \left(\frac{d\alpha}{\pi^2}\right)^{s/2+1} \Gamma\left(\frac{s+1}{2}\right) \sum_{n=1}^{\infty} \frac{1}{(n^2 + (d\alpha/\pi)^2)^{(s+1)/2}}\right)$$

$$= -\frac{1}{4}\alpha^{(-s+1)/2} \Gamma\left(\frac{s}{2}\right) \zeta(s) + \frac{1}{4}\alpha^{(-s-1)/2} \sqrt{\pi} \Gamma\left(\frac{s+1}{2}\right) \zeta(s+1)$$

$$+ \frac{1}{2}\alpha^{(s+1)/2} \sqrt{\pi} \Gamma\left(\frac{s+1}{2}\right) \sum_{d=1}^{\infty} \sum_{n=1}^{\infty} \frac{1}{(n^2\pi^2 + d^2\alpha^2)^{(s+1)/2}} \qquad (3.4)$$

$$= -\frac{1}{4}\alpha^{(-s+1)/2} \Gamma\left(\frac{s}{2}\right) \zeta(s) + \frac{1}{4}\alpha^{(-s-1)/2} \sqrt{\pi} \Gamma\left(\frac{s+1}{2}\right) \zeta(s+1)$$

$$+ \frac{1}{2}\alpha^{(-s-1)/2} \sqrt{\pi} \Gamma\left(\frac{s+1}{2}\right) \sum_{d=1}^{\infty} \sum_{n=1}^{\infty} \frac{1}{(n^2\beta^2/\pi^2 + d^2)^{(s+1)/2}}$$

$$= -\frac{1}{4}\alpha^{(-s+1)/2} \Gamma\left(\frac{s}{2}\right) \zeta(s) + \frac{1}{4}\alpha^{(-s-1)/2} \sqrt{\pi} \Gamma\left(\frac{s+1}{2}\right) \zeta(s+1)$$

$$+ \frac{1}{2}\beta^{(s+1)/2} \sqrt{\pi} \Gamma\left(\frac{s+1}{2}\right) \sum_{d=1}^{\infty} \sum_{n=1}^{\infty} \frac{1}{(n^2\beta^2 + d^2\pi^2)^{(s+1)/2}}, \qquad (3.5)$$

where we used the hypothesis $\alpha\beta = \pi^2$. By symmetry, from (3.4), for Re $s > 0$,

$$\sqrt{\beta} \sum_{n=1}^{\infty} \sigma_{-s}(n) n^{s/2} K_{s/2}(2n\beta)$$

$$= -\frac{1}{4}\beta^{(-s+1)/2}\Gamma\left(\frac{s}{2}\right)\zeta(s) + \frac{1}{4}\beta^{(-s-1)/2}\sqrt{\pi}\Gamma\left(\frac{s+1}{2}\right)\zeta(s+1)$$

$$+ \frac{1}{2}\beta^{(s+1)/2}\sqrt{\pi}\Gamma\left(\frac{s+1}{2}\right)\sum_{d=1}^{\infty}\sum_{n=1}^{\infty}\frac{1}{(n^2\pi^2 + d^2\beta^2)^{(s+1)/2}}. \quad (3.6)$$

Reversing the roles of the summation variables d and n in (3.6), subtracting (3.6) from (3.5), and rearranging slightly, we deduce that

$$\sqrt{\alpha}\sum_{n=1}^{\infty}\sigma_{-s}(n)n^{s/2}K_{s/2}(2n\alpha) - \sqrt{\beta}\sum_{n=1}^{\infty}\sigma_{-s}(n)n^{s/2}K_{s/2}(2n\beta)$$

$$= -\frac{1}{4}\alpha^{(-s+1)/2}\Gamma\left(\frac{s}{2}\right)\zeta(s) + \frac{1}{4}\alpha^{(-s-1)/2}\sqrt{\pi}\Gamma\left(\frac{s+1}{2}\right)\zeta(s+1)$$

$$+ \frac{1}{4}\beta^{(-s+1)/2}\Gamma\left(\frac{s}{2}\right)\zeta(s) - \frac{1}{4}\beta^{(-s-1)/2}\sqrt{\pi}\Gamma\left(\frac{s+1}{2}\right)\zeta(s+1). \quad (3.7)$$

On the other hand, using the functional equation (1.4) of $\zeta(s)$ and the fact that $\alpha\beta = \pi^2$, we find that

$$\frac{1}{4}\alpha^{(-s-1)/2}\sqrt{\pi}\Gamma\left(\frac{s+1}{2}\right)\zeta(s+1) = \frac{1}{4}\alpha^{(-s-1)/2}\sqrt{\pi}\pi^{s+1/2}\Gamma\left(-\frac{s}{2}\right)\zeta(-s)$$

$$= \frac{1}{4}\alpha^{(-s-1)/2}(\alpha\beta)^{(s+1)/2}\Gamma\left(-\frac{s}{2}\right)\zeta(-s)$$

$$= \frac{1}{4}\beta^{(s+1)/2}\Gamma\left(-\frac{s}{2}\right)\zeta(-s). \quad (3.8)$$

Substituting (3.8) and its analogue with the roles of α and β reversed into (3.7), we find that

$$\sqrt{\alpha}\sum_{n=1}^{\infty}\sigma_{-s}(n)n^{s/2}K_{s/2}(2n\alpha) - \sqrt{\beta}\sum_{n=1}^{\infty}\sigma_{-s}(n)n^{s/2}K_{s/2}(2n\beta)$$

$$= -\frac{1}{4}\alpha^{(-s+1)/2}\Gamma\left(\frac{s}{2}\right)\zeta(s) + \frac{1}{4}\beta^{(s+1)/2}\Gamma\left(-\frac{s}{2}\right)\zeta(-s)$$

$$+ \frac{1}{4}\beta^{(-s+1)/2}\Gamma\left(\frac{s}{2}\right)\zeta(s) - \frac{1}{4}\alpha^{(s+1)/2}\Gamma\left(-\frac{s}{2}\right)\zeta(-s). \quad (3.9)$$

The identity (3.9) is simply a rearrangement of (3.1), and so the proof of (3.1) is complete for Re $s > 0$. By analytic continuation, (3.1) is valid for all complex numbers s. □

Since $K_s(z) = K_{-s}(z)$ [37, p. 79, equation (8)], we see that (1.5) is invariant under the replacement of s by $-s$.

Ramanujan completes page 253 with two corollaries, which we now state and prove.

Entry 3.3 (p. 253). *Let α and β be positive numbers such that $\alpha\beta = \pi^2$. Then*

$$\sum_{n=1}^{\infty} \sigma_{-1}(n)e^{-2n\alpha} - \sum_{n=1}^{\infty} \sigma_{-1}(n)e^{-2n\beta} = \frac{\beta - \alpha}{12} + \frac{1}{4}\log\frac{\alpha}{\beta}. \tag{3.10}$$

Proof. Let $s = 1$ in Theorem 1.1. From (2.2),

$$\sqrt{\alpha n}\,K_{1/2}(2n\alpha) = \frac{1}{2}\sqrt{\pi}\,e^{-2n\alpha}. \tag{3.11}$$

Using (3.11), the values $\Gamma(-\frac{1}{2}) = -2\Gamma(\frac{1}{2}) = -2\sqrt{\pi}$ and $\zeta(-1) = -\frac{1}{12}$ [34, p. 19], and the Laurent expansion of $\zeta(s)$ about $s = 1$ [34, p. 16, equation (2.1.16)] in (1.5), we find that

$$\sum_{n=1}^{\infty} \sigma_{-1}(n)e^{-2n\alpha} - \sum_{n=1}^{\infty} \sigma_{-1}(n)e^{-2n\beta} - \frac{\beta - \alpha}{12}$$

$$= \frac{1}{2\sqrt{\pi}}\lim_{s \to 1}\Gamma\left(\frac{s}{2}\right)\zeta(s)\{\beta^{(1-s)/2} - \alpha^{(1-s)/2}\}$$

$$= \frac{1}{2}\lim_{s \to 1}\left(\frac{1}{s-1} + \gamma + \cdots\right)$$

$$\times \left(\left\{1 - \frac{s-1}{2}\log\beta + \cdots\right\} - \left\{1 - \frac{s-1}{2}\log\alpha + \cdots\right\}\right)$$

$$= \frac{1}{4}\log\frac{\alpha}{\beta}. \tag{3.12}$$

We easily see that (3.12) is equivalent to (3.10), and so the proof is complete. $\qquad\square$

Entry 3.3 is equivalent to the identity

$$\sum_{m=1}^{\infty} \frac{1}{m(e^{2m\alpha} - 1)} - \sum_{m=1}^{\infty} \frac{1}{m(e^{2m\beta} - 1)} = \frac{\beta - \alpha}{12} + \frac{1}{4}\log\frac{\alpha}{\beta}. \tag{3.13}$$

To see this, expand the summands in (3.13) in geometric series and collect together all terms with the same exponents in the resulting double series. The formula (3.13) (or (3.10)) is equivalent to the transformation formula for the logarithm of the Dedekind eta function. Ramanujan stated (3.13) twice in his second notebook [27], namely as Corollary (ii) in Section 8 of Chapter 14 [6, p. 256] and as Entry 27(iii) in Chapter 16 [7, p. 43]. He also stated (3.13) in an unpublished manuscript on infinite series reproduced along with Ramanujan's lost notebook [28], in particular, as formula (29) on page 320 of [28]. (See also [8, p. 65, Entry 3.5].)

We next demonstrate that Koshliakov's formula (1.1) is a corollary of Entry 3.1. Our proof is a detailed explication of that of Guinand [15].

Entry 3.4 (p. 253). *Let α and β denote positive numbers such that $\alpha\beta = \pi^2$. Then, if γ denotes Euler's constant,*

$$\sqrt{\alpha}\left(\frac{1}{4}\gamma - \frac{1}{4}\log(4\beta) + \sum_{n=1}^{\infty} d(n)K_0(2n\alpha)\right)$$

$$= \sqrt{\beta}\left(\frac{1}{4}\gamma - \frac{1}{4}\log(4\alpha) + \sum_{n=1}^{\infty} d(n)K_0(2n\beta)\right). \tag{3.14}$$

Proof. In order to let $s \to 0$ in Theorem 1.1, we need the well-known Laurent expansions [14, p. 944, formula 8.321, no. 1]

$$\Gamma(s) = \frac{1}{s} - \gamma + \cdots \tag{3.15}$$

and [34, pp. 19–20, equations (2.4.3), (2.4.5)]

$$\zeta(s) = -\frac{1}{2} - \frac{1}{2}\log(2\pi)s + \cdots . \tag{3.16}$$

Hence, letting $s \to 0$ in (1.5) and using (3.15) and (3.16), we find that

$$\sqrt{\alpha}\sum_{n=1}^{\infty} d(n)K_0(2n\alpha) - \sqrt{\beta}\sum_{n=1}^{\infty} d(n)K_0(2n\beta)$$

$$= \frac{1}{4}\lim_{s\to 0}\left(\left\{\left(\frac{1}{s/2} - \gamma + \cdots\right)\left(-\frac{1}{2} - \frac{1}{2}\log(2\pi)s + \cdots\right)\right.\right.$$

$$\times \left.\left(\sqrt{\beta}\left\{1 - \frac{1}{2}s\log\beta + \cdots\right\} - \sqrt{\alpha}\left\{1 - \frac{1}{2}s\log\alpha + \cdots\right\}\right)\right\}$$

$$+ \left\{\left(\frac{1}{-s/2} - \gamma + \cdots\right)\left(-\frac{1}{2} + \frac{1}{2}\log(2\pi)s + \cdots\right)\right.$$

$$\times \left.\left.\left(\sqrt{\beta}\left\{1 + \frac{1}{2}s\log\beta + \cdots\right\} - \sqrt{\alpha}\left\{1 + \frac{1}{2}s\log\alpha + \cdots\right\}\right)\right\}\right)$$

$$= \frac{1}{4}\gamma(\sqrt{\beta} - \sqrt{\alpha}) - \frac{1}{2}\log(2\pi)(\sqrt{\beta} - \sqrt{\alpha}) + \frac{1}{4}(\sqrt{\beta}\log\beta - \sqrt{\alpha}\log\alpha)$$

$$= \frac{1}{4}\gamma(\sqrt{\beta} - \sqrt{\alpha}) - \frac{1}{4}\log(4\alpha\beta)(\sqrt{\beta} - \sqrt{\alpha}) + \frac{1}{4}(\sqrt{\beta}\log\beta - \sqrt{\alpha}\log\alpha), \tag{3.17}$$

where in the last step we used the equality $\alpha\beta = \pi^2$. A simplification and rearrangement of (3.17) yields (3.14) to complete the proof. $\square$

4 Kindred Formulas on Page 254 of the Lost Notebook

Entry 4.1 (p. 254). *If* $a > 0$,

$$\int_0^\infty \frac{dx}{x(e^{2\pi x} - 1)(e^{2\pi a/x} - 1)} = 2 \sum_{n=1}^\infty d(n) K_0(4\pi \sqrt{an})$$

$$= \frac{a}{\pi^2} \sum_{n=1}^\infty \frac{d(n) \log(a/n)}{a^2 - n^2} - \frac{1}{2}\gamma - \left(\frac{1}{4} + \frac{1}{4\pi^2 a}\right) \log a - \frac{\log(2\pi)}{2\pi^2 a}. \tag{4.1}$$

Proof. Expanding the integrand in geometric series, we find that

$$\int_0^\infty \frac{dx}{x(e^{2\pi x} - 1)(e^{2\pi a/x} - 1)} = \sum_{m=1}^\infty \sum_{k=1}^\infty \int_0^\infty \frac{1}{x} e^{-2\pi(mx + ak/x)} dx$$

$$= \sum_{m=1}^\infty \sum_{k=1}^\infty \int_0^\infty \frac{1}{u} e^{-2\pi(u + akm/u)} du$$

$$= \sum_{n=1}^\infty d(n) \int_0^\infty \frac{1}{u} e^{-2\pi(u + an/u)} du$$

$$= 2 \sum_{n=1}^\infty d(n) K_0(4\pi \sqrt{an}),$$

which proves the first part of (4.1).

The second identity in (4.1) was actually first proved in print in 1966 by Soni [32]. Her proof is short, depends on Koshliakov's formula (1.1), and uses the integral evaluations (2.3) with $\nu = 0$ and (2.4). We use her idea to prove the second major claim of Ramanujan on page 254.   □

In contrast to the claims on the top and bottom thirds of page 254, the one claim in the middle of page 254 seems to be missing one element, and so we shall proceed as we think Ramanujan might have done. Proceeding as we did above and employing (2.1), we find that

$$\int_0^\infty \frac{dx}{\sqrt{x}(e^{2\pi x} - 1)(e^{2\pi a/x} - 1)} = \sum_{m=1}^\infty \sum_{k=1}^\infty \frac{1}{\sqrt{m}} \int_0^\infty \frac{1}{\sqrt{u}} e^{-2\pi(u + akm/u)} du$$

$$= \sum_{n=1}^\infty \sigma_{-1/2}(n) \int_0^\infty \frac{1}{\sqrt{u}} e^{-2\pi(u + an/u)} du$$

$$= 2 \sum_{n=1}^{\infty} \sigma_{-1/2}(n)(an)^{1/4} K_{1/2}(4\pi\sqrt{an})$$

$$= \frac{1}{\sqrt{2}} \sum_{n=1}^{\infty} \sigma_{-1/2}(n)e^{-4\pi\sqrt{an}}, \tag{4.2}$$

where we have used (2.2). Ramanujan's next claim gives an identity for the last series above, with a replaced by $a/4$.

Entry 4.2 (p. 254). *For $a > 0$,*

$$\sum_{n=1}^{\infty} \sigma_{-1/2}(n)e^{-2\pi\sqrt{an}} = Ka \sum_{n=1}^{\infty} \frac{\sigma_{-1/2}(n)}{(n+a)(\sqrt{n}+\sqrt{a})} + two\ trivial\ terms. \tag{4.3}$$

Evidently, K on the right-hand side of (4.3) represents an unspecified constant. Ramanujan does not divulge the identities of the "two trivial terms." Our calculation in (4.2) showing a discrepancy with the series on the left-hand side of (4.3) actually provides a clue that this series in (4.3) *should be replaced by the series on the right-hand side of* (4.2). We next state a corrected version of Entry 4.2 providing the identities of the constant and the "trivial" terms.

Entry 4.3 (p. 254). *If $a > 0$, then*

$$\sum_{n=1}^{\infty} \sigma_{-1/2}(n)e^{-4\pi\sqrt{an}} - \frac{a}{\pi} \sum_{n=1}^{\infty} \frac{\sigma_{-1/2}(n)}{(n+a)(\sqrt{n}+\sqrt{a})}$$

$$= \frac{1}{2}\zeta\left(\frac{1}{2}\right)\left(\frac{1}{\pi\sqrt{a}} - 1\right) + \frac{1}{2}\zeta\left(-\frac{1}{2}\right)\left(4\pi\sqrt{a} - \frac{1}{\pi a}\right). \tag{4.4}$$

Proof. In (1.5), set $s = \frac{1}{2}$ and $\alpha = x$, so that $\beta = \pi^2/x$. Then,

$$\sqrt{x} \sum_{n=1}^{\infty} \sigma_{-1/2}(n)n^{1/4} K_{1/4}(2nx) - \frac{\pi}{\sqrt{x}} \sum_{n=1}^{\infty} \sigma_{-1/2}(n)n^{1/4} K_{1/4}(2n\pi^2/x)$$

$$= \frac{1}{4}\Gamma\left(\frac{1}{4}\right)\zeta\left(\frac{1}{2}\right)\left(\frac{\sqrt{\pi}}{x^{1/4}} - x^{1/4}\right) + \frac{1}{4}\Gamma\left(-\frac{1}{4}\right)\zeta\left(-\frac{1}{2}\right)\left(\frac{\pi^{3/2}}{x^{3/4}} - x^{3/4}\right). \tag{4.5}$$

Multiply both sides of (4.5) by

$$\frac{1}{x^{5/2}} K_{1/4}(2a\pi^2/x)$$

and integrate over $(0, \infty)$. Inverting the order of summation and integration by absolute convergence, we find that

$$\sum_{n=1}^{\infty} \sigma_{-1/2}(n)n^{1/4} \int_0^{\infty} \frac{1}{x^2} K_{1/4}(2nx) K_{1/4}(2a\pi^2/x)dx$$

$$- \pi \sum_{n=1}^{\infty} \sigma_{-1/2}(n)n^{1/4} \int_0^{\infty} \frac{1}{x^3} K_{1/4}(2n\pi^2/x) K_{1/4}(2a\pi^2/x)dx$$

$$= \frac{1}{4}\Gamma\left(\frac{1}{4}\right)\zeta\left(\frac{1}{2}\right)(\sqrt{\pi}I_3 - I_1) + \frac{1}{4}\Gamma\left(-\frac{1}{4}\right)\zeta\left(-\frac{1}{2}\right)(\pi^{3/2}I_5 - I_{-1}), \quad (4.6)$$

where

$$I_j = \int_0^{\infty} u^{j/4} K_{1/4}(2a\pi^2 u)du, \qquad (4.7)$$

and where, to obtain the four integrals on the right-hand side of (4.6), we made the change of variable $x = 1/u$ in each one.

We examine each of the six integrals in (4.6) in turn. First, using (2.3) and (2.2), we find that

$$\int_0^{\infty} \frac{1}{x^2} K_{1/4}(2nx) K_{1/4}(2a\pi^2/x)dx = \frac{1}{2a\pi} K_{1/2}(4\pi\sqrt{an})$$

$$= \frac{1}{4\sqrt{2}a^{5/4}n^{1/4}\pi} e^{-4\pi\sqrt{an}}. \qquad (4.8)$$

Second, making the change of variable $u = \pi^2/x$ and using (2.5), we deduce that

$$\int_0^{\infty} \frac{1}{x^3} K_{1/4}(2n\pi^2/x) K_{1/4}(2a\pi^2/x)dx = \frac{1}{\pi^4} \int_0^{\infty} u K_{1/4}(2nu) K_{1/4}(2au)du$$

$$= \frac{1}{\pi^4} \frac{\pi(4na)^{-1/4}(\sqrt{2n} - \sqrt{2a})}{2\sin(\pi/4)(4n^2 - 4a^2)}$$

$$= \frac{\sqrt{2}(an)^{-1/4}}{8\pi^3(n+a)(\sqrt{n}+\sqrt{a})}. \qquad (4.9)$$

In our calculations of I_j, $j = 3, 1, 5, -1$, we employ (2.6). Thus,

$$I_3 = 2^{-1/4}(2a\pi^2)^{-7/4}\Gamma(1)\Gamma\left(\frac{3}{4}\right) = \frac{1}{4a^{7/4}\pi^{7/2}}\Gamma\left(\frac{3}{4}\right), \qquad (4.10)$$

$$I_1 = 2^{-3/4}(2a\pi^2)^{-5/4}\Gamma\left(\frac{3}{4}\right)\Gamma\left(\frac{1}{2}\right) = \frac{1}{4a^{5/4}\pi^2}\Gamma\left(\frac{3}{4}\right), \qquad (4.11)$$

$$I_5 = 2^{1/4}(2a\pi^2)^{-9/4}\Gamma\left(\frac{5}{4}\right)\Gamma(1) = \frac{1}{4a^{9/4}\pi^{9/2}}\Gamma\left(\frac{5}{4}\right), \qquad (4.12)$$

$$I_{-1} = 2^{-5/4}(2a\pi^2)^{-3/4}\Gamma\left(\frac{1}{2}\right)\Gamma\left(\frac{1}{4}\right) = \frac{1}{4a^{3/4}\pi}\Gamma\left(\frac{1}{4}\right). \qquad (4.13)$$

Using (4.8)–(4.13) in (4.6) and making frequent use of the reflection formula

$$\Gamma(z)\Gamma(1-z) = \frac{\pi}{\sin(\pi z)},$$

we deduce that

$$\frac{1}{4\sqrt{2}a^{5/4}\pi} \sum_{n=1}^{\infty} \sigma_{-1/2}(n)e^{-4\pi\sqrt{an}} - \frac{1}{4\sqrt{2}a^{1/4}\pi^2} \sum_{n=1}^{\infty} \frac{\sigma_{-1/2}(n)}{(n+a)(\sqrt{n}+\sqrt{a})}$$

$$= \frac{\sqrt{2}}{16} \zeta\left(\frac{1}{2}\right)\left(\frac{1}{a^{7/4}\pi^2} - \frac{1}{a^{5/4}\pi}\right) + \frac{\sqrt{2}}{16} \zeta\left(-\frac{1}{2}\right)\left(-\frac{1}{a^{9/4}\pi^2} + \frac{4}{a^{3/4}}\right).$$

$$(4.14)$$

If we multiply both sides of (4.14) by $4\sqrt{2}a^{5/4}\pi$ and rearrange slightly, we obtain (4.4) to complete the proof. □

We record the last two results on page 254 as Ramanujan wrote them, except that we express the results in terms of Bessel functions. The constant K and the "two trivial terms" are not necessarily the same as they are in Entry 4.2.

Entry 4.4 (p. 254). *If $a > 0$, then*

$$\int_0^{\infty} \frac{dx}{(e^{2\pi x}-1)(e^{2\pi a/x}-1)} = 2\sqrt{a}\sum_{n=1}^{\infty} \sigma_{-1}(n)\sqrt{n}K_1(4\pi\sqrt{an}) \qquad (4.15)$$

$$= Ka^2 \sum_{n=1}^{\infty} \frac{\sigma_{-1}(n)}{n(n+a)} + \textit{two trivial terms.} \qquad (4.16)$$

Proof. We prove (4.15). Expanding the integrand in geometric series, setting $mx = u$, and invoking (2.1), we find that

$$\int_0^{\infty} \frac{dx}{(e^{2\pi x}-1)(e^{2\pi a/x}-1)} = \sum_{m=1}^{\infty}\sum_{k=1}^{\infty} \frac{1}{m} \int_0^{\infty} e^{-2\pi(u+akm/u)}du$$

$$= \sum_{n=1}^{\infty} \sigma_{-1}(n) \int_0^{\infty} e^{-2\pi(u+an/u)}du$$

$$= 2\sqrt{a}\sum_{n=1}^{\infty} \sigma_{-1}(n)\sqrt{n}K_1(4\pi\sqrt{an}).$$

□

Lastly, we provide and prove a more precise version of (4.16) giving the identities of the missing terms.

Entry 4.5 (p. 254). *If $a > 0$ and γ denotes Euler's constant, then*

$$2\sqrt{a}\sum_{n=1}^{\infty}\sigma_{-1}(n)\sqrt{n}K_1(4\pi\sqrt{an})$$

$$= -\frac{a^2}{2\pi}\sum_{n=1}^{\infty}\frac{\sigma_{-1}(n)}{n(n+a)} + \frac{a}{2\pi}((\log a + \gamma)\zeta(2) + \zeta'(2))$$

$$+ \frac{1}{4\pi}(\log 2a\pi + \gamma) + \frac{1}{48a\pi}. \tag{4.17}$$

Proof. In (3.10), set $\alpha = x$, so that $\beta = \pi^2/x$. Recalling (2.2), we find that

$$\frac{2}{\sqrt{\pi}}\sum_{n=1}^{\infty}\sigma_{-1}(n)\sqrt{nx}K_{1/2}(2nx)$$

$$= \left(\sum_{n=1}^{\infty}\sigma_{-1}(n)e^{-2n\pi^2/x} - \frac{x}{12}\right) + \frac{1}{2}\log\frac{x}{\pi} + \frac{\pi^2}{12x}$$

$$=: I_1 + I_2 + I_3. \tag{4.18}$$

Next, multiply both sides of (4.18) by

$$\frac{1}{x^{5/2}}K_{1/2}(2a\pi^2/x)$$

and integrate over $(0, \infty)$.

Consider first the series arising on the left-hand side of (4.18). Inverting the order of summation and integration on the left-hand side by absolute convergence, we arrive at

$$\frac{2}{\sqrt{\pi}}\sum_{n=1}^{\infty}\sigma_{-1}(n)\sqrt{n}\int_0^{\infty}\frac{1}{x^2}K_{1/2}(2nx)K_{1/2}(2a\pi^2/x)dx$$

$$= \frac{1}{a\pi^{3/2}}\sum_{n=1}^{\infty}\sigma_{-1}(n)\sqrt{n}K_1(4\pi\sqrt{an}), \tag{4.19}$$

where we have employed (2.3).

Second, the contribution from I_3 in (4.18) is given by

$$\frac{\pi^2}{12}\int_0^{\infty}x^{-7/2}K_{1/2}(2a\pi^2/x)dx = \frac{\pi^2}{12}\int_0^{\infty}u^{3/2}K_{1/2}(2a\pi^2u)du = \frac{1}{96a^{5/2}\pi^{5/2}}, \tag{4.20}$$

where we used (2.6) in the last step with $\mu = 3/2$, $\nu = 1/2$, and a replaced by $2a\pi^2$.

Third, using (2.2), we find that the contribution from I_2 in (4.18) is equal to

$$\frac{1}{2}\int_0^\infty x^{-5/2}\log(x/\pi)K_{1/2}(2a\pi^2/x)dx$$

$$=\frac{1}{4\sqrt{a\pi}}\int_0^\infty x^{-2}\log(x/\pi)e^{-2a\pi^2/x}dx$$

$$=\frac{1}{8a^{3/2}\pi^{5/2}}\int_0^\infty \log(2a\pi/u)e^{-u}du$$

$$=\frac{1}{8a^{3/2}\pi^{5/2}}\left\{\int_0^\infty e^{-u}\log(2a\pi)du - \int_0^\infty e^{-u}\log u\, du\right\}$$

$$=\frac{1}{8a^{3/2}\pi^{5/2}}\left\{\log(2a\pi) - \int_0^\infty e^{-u}\log u\, du\right\}$$

$$=\frac{1}{8a^{3/2}\pi^{5/2}}\{\log(2a\pi)+\gamma\}, \tag{4.21}$$

since [14, p. 602, formula 4.331, no. 1]

$$\gamma = -\int_0^\infty e^{-u}\log u\, du.$$

Finally, the contribution from I_1 in (4.18) is given by

$$J := \int_0^\infty\left(\sum_{n=1}^\infty\sigma_{-1}(n)e^{-2n\pi^2/x} - \frac{1}{12}x\right)x^{-5/2}K_{1/2}(2a\pi^2/x)dx. \tag{4.22}$$

Note that $\zeta(2) = \pi^2/6$. Thus, we can write

$$\sum_{n=1}^\infty\sigma_{-1}(n)e^{-2n\pi^2/x} - \frac{1}{12}x = \sum_{n=1}^\infty\sum_{d|n}\frac{1}{d}e^{-2n\pi^2/x} - \frac{1}{12}x$$

$$=\sum_{d=1}^\infty\sum_{m=1}^\infty\frac{1}{d}e^{-2md\pi^2/x} - \frac{1}{12}x$$

$$=\sum_{d=1}^\infty\frac{1}{d}\frac{1}{e^{2d\pi^2/x}-1} - \left(\sum_{n=1}^\infty\frac{1}{n^2}\right)\frac{x}{2\pi^2}. \tag{4.23}$$

Using (4.23) and (2.2) in (4.22), we see that

$$J = \int_0^\infty\left(\sum_{n=1}^\infty\frac{1}{n}\frac{1}{e^{2n\pi^2/x}-1} - \left(\sum_{n=1}^\infty\frac{1}{n^2}\right)\frac{x}{2\pi^2}\right)\frac{1}{2\sqrt{a\pi}}e^{-2a\pi^2/x}\frac{dx}{x^2}$$

$$=\frac{1}{2\sqrt{a\pi}}\int_0^\infty\sum_{n=1}^\infty\frac{1}{n}\left(\frac{1}{e^{2n\pi^2/x}-1} - \frac{1}{2n\pi^2/x}\right)e^{-2a\pi^2/x}\frac{dx}{x^2}. \tag{4.24}$$

Since for $z > 0$,

$$\frac{1}{e^z - 1} - \frac{1}{z} < 0,$$

we can change the order of summation and integration by the monotone convergence theorem. Hence,

$$J = \frac{1}{2\sqrt{a\pi}} \sum_{n=1}^{\infty} \frac{1}{n} \int_0^{\infty} \left(\frac{1}{e^{2n\pi^2/x} - 1} - \frac{1}{2n\pi^2/x} \right) e^{-2a\pi^2/x} \frac{dx}{x^2}$$

$$= \frac{1}{4\sqrt{a}\pi^{5/2}} \sum_{n=1}^{\infty} \frac{1}{n^2} \int_0^{\infty} \left(\frac{1}{e^u - 1} - \frac{1}{u} \right) e^{-au/n} du. \tag{4.25}$$

Consider now two different expressions for the logarithmic derivative of the gamma function, namely [14, p. 952, formula 8.362, no. 1; formula 8.361, no. 8],

$$\frac{\Gamma'(z)}{\Gamma(z)} = -\gamma - \frac{1}{z} + \sum_{n=1}^{\infty} \frac{z}{n(n+z)}$$

$$= \log z - \frac{1}{z} - \int_0^{\infty} \left(\frac{1}{e^t - 1} - \frac{1}{t} \right) e^{-tz} dt,$$

where Re $z > 0$. Hence,

$$\int_0^{\infty} \left(\frac{1}{e^u - 1} - \frac{1}{u} \right) e^{-au/n} du = \log(a/n) + \gamma - \sum_{m=1}^{\infty} \frac{a}{m(mn + a)}. \tag{4.26}$$

Putting (4.26) in (4.25), we find that

$$J = \frac{1}{4a^{1/2}\pi^{5/2}} \sum_{n=1}^{\infty} \frac{1}{n^2} \left(\log(a/n) + \gamma - \sum_{m=1}^{\infty} \frac{a}{m(mn + a)} \right)$$

$$= \frac{1}{4a^{1/2}\pi^{5/2}} \left((\log a + \gamma)\zeta(2) - \sum_{n=1}^{\infty} \frac{\log n}{n^2} - \sum_{n=1}^{\infty} \sum_{m=1}^{\infty} \frac{a}{n^2 m(mn + a)} \right)$$

$$= \frac{1}{4a^{1/2}\pi^{5/2}} \left((\log a + \gamma)\zeta(2) + \zeta'(2) - a \sum_{n=1}^{\infty} \frac{\sigma_{-1}(n)}{n(n + a)} \right)$$

$$= -\frac{a^{1/2}}{4\pi^{5/2}} \sum_{n=1}^{\infty} \frac{\sigma_{-1}(n)}{n(n + a)} + \frac{1}{4a^{1/2}\pi^{5/2}} ((\log a + \gamma)\zeta(2) + \zeta'(2)). \tag{4.27}$$

We now combine all our calculations that arose from (4.18), namely, (4.19), (4.20), (4.21), (4.22), and (4.27), to deduce that

$$\frac{1}{a\pi^{3/2}} \sum_{n=1}^{\infty} \sigma_{-1}(n)\sqrt{n}K_1(4\pi\sqrt{an})$$

$$= \frac{1}{96a^{5/2}\pi^{5/2}} + \frac{1}{8a^{3/2}\pi^{5/2}}\{\log(2a\pi) + \gamma\}$$

$$- \frac{a^{1/2}}{4\pi^{5/2}} \sum_{n=1}^{\infty} \frac{\sigma_{-1}(n)}{n(n+a)} + \frac{1}{4a^{1/2}\pi^{5/2}}((\log a + \gamma)\zeta(2) + \zeta'(2)). \qquad (4.28)$$

Finally multiply both sides of (4.28) by $2\pi^{3/2}a^{3/2}$ to deduce (4.17) and complete the proof. $\qquad\Box$

5 Analogues of Guinand's Formula

Extensive analogues of Guinand's formula in Theorem 3.1 and Watson's lemma (Lemma 3.2) have been derived by Berndt [3]. To state these theorems completely, we need a lengthy definition. Throughout this section, $s = \sigma + it$, with σ and t both real.

Definition 5.1. Let λ_n and μ_n, $1 \le n < \infty$, be two sequences of positive numbers strictly increasing to ∞, and let $a(n)$ and $b(n)$, $1 \le n < \infty$, be two sequences of complex numbers not identically zero. Consider the functions $\varphi(s)$ and $\psi(s)$ representable as Dirichlet series

$$\varphi(s) = \sum_{n=1}^{\infty} a(n)\lambda_n^{-s} \quad \text{and} \quad \psi(s) = \sum_{n=1}^{\infty} b(n)\mu_n^{-s}$$

with finite abscissas of absolute convergence σ_a and σ_a^*, respectively. For any positive integer N, let

$$\Delta(s) = \prod_{k=1}^{N} \Gamma(\alpha_k s + \beta_k),$$

where $\alpha_k > 0$ and β_k is complex, $1 \le k \le N$. If r is real, we say that φ and ψ satisfy the functional equation

$$\Delta(s)\varphi(s) = \Delta(r-s)\psi(r-s) \qquad (5.1)$$

if there exists in the s-plane a domain $\mathcal{D}$, the exterior of a compact set S, such that in $\mathcal{D}$ a holomorphic function χ exists with the following properties:

(i) $\chi(s) = \Delta(s)\varphi(s), \quad \sigma > \sigma_a, \quad \chi(s) = \Delta(r-s)\psi(r-s), \quad \sigma < r - \sigma_a*;$

(ii) $\lim_{|t|\to\infty} \chi(\sigma + it) = 0,$

   uniformly in every interval $-\infty < \sigma_1 \le \sigma \le \sigma_2 < \infty$.

In particular, if $\Delta(s) = \Gamma(s)$, then (5.1) reduces to Hecke's functional equation

$$\Gamma(s)\varphi(s) = \Gamma(r-s)\psi(r-s). \tag{5.2}$$

We now give an analogue of Lemma 3.2; a proof can be found in [3, p. 342, Theorem 8.1].

Theorem 5.2. *Suppose that $\varphi(s)$ satisfies Definition 5.1 and (5.2). Define for $a > 0$ and $\sigma > \sigma_a$,*

$$\varphi(s, a) = \sum_{n=1}^{\infty} a(n)(\lambda_n + a)^{-s}.$$

Let $\mathbb{D}$ be a domain on which

$$\sum_{n=1}^{\infty} b(n) K_{s-r}(2\sqrt{\mu_n a})\mu_n^{(s-r)/2}$$

converges uniformly. Suppose that the singularities of $\chi(w)$ are at most poles. Let $R(s, a)$ denote the sum of the residues of $\chi(w)\Gamma(s-w)a^{w-s}$ at the poles of $\chi(w)$. Then, if $s \in \mathbb{D}$,

$$\Gamma(s)\varphi(s, a) = 2 \sum_{n=1}^{\infty} b(n) \left(\frac{\mu_n}{a}\right)^{(s-r)/2} K_{s-r}(2\sqrt{\mu_n a}) + R(s, a). \tag{5.3}$$

Conversely, if $\varphi(s, a)$ satisfies (5.3), then φ satisfies (5.2).

Lastly, we state an analogue of Guinand's formula, Entry 3.1; see [3, p. 343, Theorem 9.1] for a proof.

Theorem 5.3. *Suppose that $\varphi(s)$ satisfies Definition 5.1 and (5.2). Define, for $x > 0$,*

$$P(x) := \frac{1}{2\pi i} \int_C \chi(s) x^{-s} ds,$$

where C is a curve, or curves, containing S from Definition 5.1 on its interior. If s is any complex number and $\operatorname{Re} a, \operatorname{Re} b > 0$, then

$$2 \sum_{n=1}^{\infty} a(n) \left(\frac{b}{\lambda_n + a}\right)^{s/2} K_s \left(2\sqrt{(\lambda_n + a)b}\right)$$

$$= \int_0^{\infty} x^{s-1} e^{-ax - b/x} P(x) dx$$

$$+ 2 \sum_{n=1}^{\infty} b(n) \left(\frac{a}{\mu_n + b}\right)^{(r-s)/2} K_{r-s} \left(2\sqrt{(\mu_n + b)a}\right).$$

Example 1. Let $r_k(n)$ denote the number of representations of the positive integer n as a sum of k squares, where $k \geq 2$. Then

$$\varphi(s) := \zeta_k(s) = \sum_{n=1}^{\infty} r_k(n) n^{-s}, \quad \sigma > 1,$$

denotes the Epstein zeta function associated with the quadratic form $n_1^2 + n_2^2 + \cdots + n_k^2$. It is well known that $\zeta_k(s)$ satisfies the functional equation [10, p. 18]

$$\pi^{-s} \Gamma(s) \zeta_k(s) = \pi^{-(k/2-s)} \Gamma\left(\frac{1}{2}k - s\right) \zeta_k\left(\frac{1}{2}k - s\right).$$

Thus, $a(n) = b(n) = r_k(n)$, $\lambda_n = \mu_n = \pi n$, and $r = \frac{1}{2}k$. It is also known [10, p. 19] that $P(x) = -1 + \pi^{k/2} x^{-k/2}$. Hence, from (5.3), for any $s \in \mathbb{C}$,

$$\int_0^{\infty} x^{s-1} e^{-ax-b/x} P(x) dx$$

$$= \int_0^{\infty} e^{-ax-b/x} (-x^{s-1} + \pi^{k/2} x^{s-k/2-1}) dx$$

$$= -2 \left(\frac{b}{a}\right)^{s/2} K_s(2\sqrt{ab}) + 2 \left(\frac{b}{a}\right)^{(s-k/2)/2} K_{s-k/2}(2\sqrt{ab}), \qquad (5.4)$$

where we have made two applications of (2.1). Define $r_k(0) = 1$. In Theorem 5.3, replace a and b by πa and πb, respectively, and use (5.4) to deduce that for all complex numbers s,

$$\sum_{n=0}^{\infty} r_k(n) \left(\frac{b}{n+a}\right)^{s/2} K_s\left(2\pi\sqrt{(n+a)b}\right)$$

$$= \sum_{n=0}^{\infty} r_k(n) \left(\frac{a}{n+b}\right)^{(k/2-s)/2} K_{k/2-s}\left(2\pi\sqrt{(n+b)a}\right). \qquad (5.5)$$

In particular, if $s = \frac{1}{2}$ and $k = 2$, then on the use of (2.2), we see that (5.5) reduces to the identity

$$\sum_{n=0}^{\infty} \frac{r_2(n)}{\sqrt{n+a}} e^{-2\pi\sqrt{(n+a)b}} = \sum_{n=0}^{\infty} \frac{r_2(n)}{\sqrt{n+b}} e^{-2\pi\sqrt{(n+b)a}}. \qquad (5.6)$$

This identity was first proved by Ramanujan and communicated to G.H. Hardy, who recorded and sketched a proof of it in his paper [16, p. 283], [17, p. 263] on the famous "circle problem." In fact, Hardy derived the corresponding identity for representations of an integer by an arbitrary positive definite quadratic form in two variables. The identity (5.6) cannot be found in Ramanujan's published papers or notebooks. Another proof of (5.6) was given by Dixon and Ferrar [11]. The symmetry of (5.5) and (5.6) in the parameters a and b is striking.

Example 2. Let $\tau(n)$ denote Ramanujan's famous τ-function, and let

$$R(s) := \sum_{n=1}^{\infty} \tau(n)n^{-s}, \quad \sigma > \frac{13}{2},$$

denote Ramanujan's Dirichlet series. Then $R(s)$ satisfies the functional equation [24]

$$(2\pi)^{-s}\Gamma(s)R(s) = (2\pi)^{-(12-s)}\Gamma(12-s)R(12-s).$$

We apply (5.3) with $\lambda_n = \mu_n = 2\pi n$ and $a(n) = b(n) = \tau(n)$, $n \geq 1$. Also [10, p. 16], [24], $P(x) \equiv 0$. Hence, if we replace a and b by $2\pi a$ and $2\pi b$, respectively, we find that for any complex number s,

$$\sum_{n=1}^{\infty} \tau(n) \left(\frac{b}{n+a}\right)^{s/2} K_s\left(4\pi\sqrt{(n+a)b}\right)$$

$$= \sum_{n=1}^{\infty} \tau(n) \left(\frac{a}{n+b}\right)^{(12-s)/2} K_{12-s}\left(4\pi\sqrt{(n+b)a}\right), \tag{5.7}$$

which indeed is a beautiful identity.

The authors are pleased to acknowledge the helpful comments of Scott Ahlgren, Carlos Moreno, and Yoshio Tanigawa.

References

1. G.E. Andrews and B.C. Berndt, *Ramanujan's Lost Notebook*, Part II, Springer, New York, 2008, to appear.
2. P.T. Bateman and E. Grosswald, *On Epstein's zeta function*, Acta Arith. **9** (1964), 365–373.
3. B.C. Berndt, *Identities involving the coefficients of a class of Dirichlet series. III.*, Trans. Amer. Math. Soc. **146** (1969), 323–348.
4. B.C. Berndt, *Identities involving the coefficients of a class of Dirichlet series. V.*, Trans. Amer. Math. Soc. **160** (1971), 139–156.
5. B.C. Berndt, *Periodic Bernoulli numbers, summmation formulas and applications*, in *Theory and Application of Special Functions*, R.A. Askey, ed., Academic Press, New York, 1975, pp. 143–189.
6. B.C. Berndt, *Ramanujan's Notebooks, Part* II, Springer–Verlag, New York, 1989.
7. B.C. Berndt, *Ramanujan's Notebooks, Part* III, Springer–Verlag, New York, 1991.
8. B.C. Berndt, *An unpublished manuscript of Ramanujan on infinite series identities*, J. Ramanujan Math. Soc. **19** (2004), 57–74.
9. B.C. Berndt and A.J. Yee, *Ramanujan's contributions to Eisenstein series, especially in his lost notebook*, in *Number-Theoretic Methods – Future Trends*, C. Jia and S. Kanemitsu, eds., Kluwer, Dordrecht, 2002, pp. 31–53; abridged version, *A survey on Eisenstein series in Ramanujan's lost notebook*, in *New Aspects of Analytic Number Theory*, Y. Tanigawa, ed., Research Institute for Mathematical Sciences, Kyoto University, Kyoto, 2002, pp. 130–141.

10. K. Chandrasekharan and R. Narasimhan, *Hecke's functional equation and arithmetical identities*, Ann. Math. (2) **74** (1961), 1–23.

11. A.L. Dixon and W.L. Ferrar, *Some summations over the lattice points of a circle* (I), Quart. J. Math. (Oxford), **5** (1934), 48–63.

12. A.L. Dixon and W.L. Ferrar, *On the summation formulae of Voronoï and Poisson*, Quart. J. Math. (Oxford) **8** (1937), 66–74.

13. W.L. Ferrar, *Some solutions of the equation* $F(t) = F(t^{-1})$, J. London Math. Soc. **11** (1936), 99–103.

14. I.S. Gradshteyn and I.M. Ryzhik, eds., *Table of Integrals, Series, and Products*, 5th ed., Academic Press, San Diego, 1994.

15. A.P. Guinand, *Some rapidly convergent series for the Riemann ξ-function*, Quart. J. Math. (Oxford) **6** (1955), 156–160.

16. G.H. Hardy, *On the expression of a number as the sum of two squares*, Quart. J. Math. (Oxford) **46** (1915), 263–283.

17. G.H. Hardy, *Collected Papers*, Vol. II, Oxford University Press, Oxford, 1967.

18. S. Kanemitsu, Y. Tanigawa, H. Tsukada, and M. Yoshimoto, *On Bessel series expressions for some lattice sums*: II, J. Physics A: Mathematics and General, **37** (2004), 719–734.

19. S. Kanemitsu, Y. Tanigawa, and M. Yoshimoto, *On rapidly convergent series for the Riemann zeta-values via the modular relation*, Abh. Math. Sem. Univ. Hamburg **72** (2002), 187–206.

20. H. Kober, *Transformationsformeln gewisser Besselscher Reihen Beziehungen zu Zeta-functionen*, Math. Z. **39** (1934), 609–624.

21. N.S. Koshliakov, *On Voronoï's sum-formula*, Mess. Math. **58** (1929), 30–32.

22. T. Kubota, *Elementary Theory of Eisenstein Series*, Kodansha, Tokyo, 1973.

23. H. Maass, *Über eine neue Art von nichtanalytischen automorphen Funktionen und die Bestimmung Dirichletscher Reihen durch Funktionalgleichungen*, Math. Ann. **121** (1949), 141–183.

24. L.J. Mordell, *On Mr Ramanujan's empirical expansions of modular functions*, Proc. Cambridge Philos. Soc. **19** (1917), 117–124.

25. C.J. Moreno, *Advanced Analytic Number Theory: L-Functions*, Math. Surveys and Monographs, Vol. 115, American Mathematical Society, Providence, RI, 2005.

26. F. Oberhettinger and K.L. Soni, *On some relations which are equivalent to functional equations involving the Riemann zeta function*, Math. Z. **127** (1972), 17–34.

27. S. Ramanujan, *Notebooks* (2 volumes), Tata Institute of Fundamental Research, Bombay, 1957.

28. S. Ramanujan, *The Lost Notebook and Other Unpublished Papers*, Narosa, New Delhi, 1988.

29. A. Selberg, *Collected Papers*, Vol. I, Springer-Verlag, Berlin, 1989.

30. A. Selberg and S. Chowla, *On Epstein's zeta-function* (I), Proc. Nat. Acad. Sci. (USA) **35** (1949), 371–374.

31. A. Selberg and S. Chowla, *On Epstein's zeta-function*, J. Reine Angew. Math. **227** (1967), 86–110.

32. K. Soni, *Some relations associated with an extension of Koshliakov's formula*, Proc. Amer. Math. Soc. **17** (1966), 543–551.

33. A. Terras, *Harmonic Analysis on Symmetric Spaces and Applications* I, Springer-Verlag, New York, 1985.

34. E.C. Titchmarsh, *The Theory of the Riemann Zeta-function*, Clarendon Press, Oxford, 1951.
35. M.G. Voronoï, *Sur une fonction transcendante et ses applications à la sommation de quelques séries*, Ann. École Norm. Sup. (3) **21** (1904), 207–267, 459–533.
36. G.N. Watson, *Some self-reciprocal functions*, Quart. J. Math. (Oxford) **2** (1931), 298–309.
37. G.N. Watson, *Theory of Bessel Functions*, 2nd ed., University Press, Cambridge, 1966.
38. S.P. Zwegers, *Mock Theta Functions*, Doctoral Dissertation, Universiteit Utrecht, 2002.

3

Exploiting Symmetries: Alternating-Sign Matrices and the Weyl Character Formulas

David M. Bressoud

Macalester College, St. Paul, MN 55105
bressoud@macalester.edu

Summary. This paper illustrates some of the power and beauty of determinant evaluations, beginning with Cauchy's proof of the Vandermonde determinant evaluation, continuing through the Weyl denominator formulas and some open conjectures on alternating-sign matrices, and ending with the Izergin–Korepin determinant expansion for the six-vertex model with domain wall boundary conditions.

Key words: Vandermonde product, Weyl denominator formula, alternating-sign matrix, six-vertex model.

2000 *Mathematics Subject Classifications*: Primary 15A15; Secondary 01A60, 05A15

1 Introduction

Symmetry is one of the great organizing principles of mathematics. If the purpose of mathematics is to discover the underlying patterns of our universe, then the search for symmetry indicates where to look for these patterns. Symmetry is also a powerful tool that provides us with simple and elegant proofs of algebraic results. At the heart of any argument involving symmetry lies the full symmetric group S_n. The structure of the full symmetry group is intimately bound to the determinant. The purpose of this paper is to illustrate some of the power and beauty of determinant evaluations, beginning with Cauchy's proof of the Vandermonde determinant evaluation and ending with the Izergin–Korepin determinant expansion for the six-vertex model with domain wall boundary conditions.

In Section 2, I will describe how symmetry leads to simple proofs of the Vandermonde determinant formula, and how this approach is generalized to the Weyl denominator formulas. Sections 3 through 5 introduce alternating-sign matrices, explain their connection to the Vandermonde determinant, and describe some of the conjectures that arose between 1980 and 1995. In Section 6, I will explain the connection between alternating-sign matrices and the 6-vertex model of statistical

K. Alladi (ed.), *Surveys in Number Theory*, DOI: 10.1007/978-0-387-78510-3_3,

mechanics, sketch the proof of the determinant formula for the partition function of the six-vertex model, and summarize recent work of Kuperberg and Okada that exploits the symmetry implicit in Baxter's triangle-to-triangle relation.

2 The Vandermonde Determinant and Its Successors

The Vandermonde determinant formula was first stated in its full generality by A.-L. Cauchy in 1815 [5]:

$$
\begin{vmatrix}
x_1^{n-1} & x_2^{n-1} & \cdots & x_n^{n-1} \\
\vdots & \vdots & \ddots & \vdots \\
x_1^1 & x_2^1 & \cdots & x_n^1 \\
x_1^0 & x_2^0 & \cdots & x_n^0
\end{vmatrix}
= \sum_{\sigma \in \mathcal{S}_n} (-1)^{\mathrm{Inv}(n)} \prod_{i=1}^{n} x_i^{n-\sigma(i)}
= \prod_{1 \le i < j \le n} (x_i - x_j). \quad (1)
$$

His proof is elegant. If we define the determinant as the given sum over permutations, then the determinant is an alternating polynomial in the variables $x_1, \ldots, x_n$ (interchanging any two of the variables changes the sign of the expression). This means that if for any pair $i < j$ we have $x_i = x_j$, then the value of the polynomial is 0. This polynomial is divisible by $\prod_{i<j}(x_i - x_j)$. Since the determinant and the product have the same total degree, their ratio must be constant. Comparing coefficients of the monomial $x_1^{n-1} x_2^{n-2} \cdots x_{n-1}^1 x_n^0$ reveals that this constant is 1.

Cauchy obtained more than just the Vandermonde determinant formula. He observed that *any* alternating polynomial is divisible by this Vandermonde product, and the resulting ratio is a symmetric polynomial. In particular, he observed that if $\lambda = (\lambda_1, \lambda_2, \ldots, \lambda_n) \in \mathbb{Z}^n$ where $\lambda_1 \ge \lambda_2 \ge \cdots \ge \lambda_n \ge 0$, then

$$
\frac{
\begin{vmatrix}
x_1^{\lambda_1+n-1} & x_2^{\lambda_1+n-1} & \cdots & x_n^{\lambda_1+n-1} \\
\vdots & \vdots & \ddots & \vdots \\
x_1^{\lambda_{n-1}+1} & x_2^{\lambda_{n-1}+1} & \cdots & x_n^{\lambda_{n-1}+1} \\
x_1^{\lambda_n} & x_2^{\lambda_n} & \cdots & x_n^{\lambda_n}
\end{vmatrix}
}{
\begin{vmatrix}
x_1^{n-1} & x_2^{n-1} & \cdots & x_n^{n-1} \\
\vdots & \vdots & \ddots & \vdots \\
x_1^1 & x_2^1 & \cdots & x_n^1 \\
x_1^0 & x_2^0 & \cdots & x_n^0
\end{vmatrix}
} = s_\lambda(x_1, \ldots, x_n) \quad (2)
$$

is a symmetric polynomial.

Today we call this polynomial the *Schur function*, named for Issai Schur, who in 1917 [17], recognized it as the character of the irreducible representation of GL_n indexed by λ. As the character of this irreducible representation, it has several remarkable properties. We let e_i denote the unit vector in the ith direction, A_{n-1} the $(n-1)$-dimensional set of vectors

$$A_{n-1} = \{\pm(e_i - e_j) \mid 1 \leq i < j \leq n\},$$

A_{n-1}^+ the subset of *positive* vectors, $e_i - e_j$, $i < j$, and ρ the half-sum of the elements of A_{n-1}^+,

$$\rho = \frac{1}{2} \sum_{r \in A_{n-1}^+} r = \left(\frac{n-1}{2}, \frac{n-3}{2}, \ldots, \frac{1-n}{2}\right).$$

The symmetric group $\mathcal{S}_n$ acts on A_{n-1} by permuting the subscripts of the unit vectors e_i. When all variables in the Schur function are set equal to 1, its value is the dimension of this representation, and this dimension can be calculated by the formula

$$s_\lambda(1, 1, \ldots, 1) = \prod_{r \in A_{n-1}^+} \frac{(\rho + \lambda) \cdot r}{\rho \cdot r}. \tag{3}$$

Hermann Weyl extended this result to other classical groups in his 1939 treatise *The Classical Groups: Their Invariants and Representations* [18]. For the subgroup of isometries within GL_{2n}, the sympletic group Sp_{2n}, the character of the representation indexed by the partition λ is given by the ratio of determinants

$$\mathrm{Sp}_{2n}(\lambda; \vec{x})$$

$$= \frac{\begin{vmatrix} x_1^{\lambda_1+n} - x_1^{-\lambda_1-n} & x_2^{\lambda_1+n} - x_2^{-\lambda_1-n} & \cdots & x_n^{\lambda_1+n} - x_n^{-\lambda_1-n} \\ \vdots & \vdots & \ddots & \vdots \\ x_1^{\lambda_{n-1}+2} - x_1^{-\lambda_{n-1}-2} & x_2^{\lambda_{n-1}+2} - x_2^{-\lambda_{n-1}-2} & \cdots & x_n^{\lambda_{n-1}+2} - x_n^{-\lambda_{n-1}-2} \\ x_1^{\lambda_n+1} - x_1^{-\lambda_n-1} & x_2^{\lambda_n+1} - x_2^{-\lambda_n-1} & \cdots & x_n^{\lambda_n+1} - x_n^{-\lambda_n-1} \end{vmatrix}}{\begin{vmatrix} x_1^n - x_1^{-n} & x_2^n - x_2^{-n} & \cdots & x_n^n - x_n^{-n} \\ \vdots & \vdots & \ddots & \vdots \\ x_1^2 - x_1^{-2} & x_2^2 - x_2^{-2} & \cdots & x_n^2 - x_n^{-2} \\ x_1^1 - x_1^{-1} & x_2^1 - x_2^{-1} & \cdots & x_n^1 - x_n^{-1} \end{vmatrix}}. \tag{4}$$

This character has a similar formula for the dimension:

$$\mathrm{Sp}(\lambda; \vec{1}) = \prod_{r \in C_n^+} \frac{(\rho + \lambda) \cdot r}{\rho \cdot r}, \tag{5}$$

where C_n^+ is defined as

$$C_n^+ = \{e_i \pm e_j \mid 1 \leq i < j \leq n\} \cup \{2i \mid 1 \leq i \leq n\},$$

and ρ is the half-sum of these vectors:

$$\rho = \frac{1}{2} \sum_{r \in C_n^+} r = (n, n-1, \ldots, 1).$$

Like the Vandermonde product, the denominator of this character has a product representation with a comparably simple proof. If we multiply this determinant by the monomial $(x_1 x_2 \cdots x_n)^n$, we get an alternating polynomial that when divided by the Vandermonde product, yields a symmetric polynomial:

$$
\begin{vmatrix}
x_1^n - x_1^{-n} & x_2^n - x_2^{-n} & \cdots & x_n^n - x_n^{-n} \\
\vdots & \vdots & \ddots & \vdots \\
x_1^2 - x_1^{-2} & x_2^2 - x_2^{-2} & \cdots & x_n^2 - x_n^{-2} \\
x_1^1 - x_1^{-1} & x_2^1 - x_2^{-1} & \cdots & x_n^1 - x_n^{-1}
\end{vmatrix}
\frac{(x_1 x_2 \cdots x_n)^n}{\prod_{1 \le i < j \le n}(x_i - x_j)}.
$$

As a polynomial in x_1, the degree is $n+1$. By inspection, we see that this polynomial is zero when $x_1 = \pm 1$ and when $x_1 = x_j^{-1}$, $j \ge 2$. By symmetry, it must equal

$$
\prod_{j=1}^{n}(x_j^2 - 1) \prod_{1 \le i < j \le n}(x_i x_j - 1),
$$

up to a constant factor that is easily checked to be 1. It follows that

$$
\begin{vmatrix}
x_1^n - x_1^{-n} & x_2^n - x_2^{-n} & \cdots & x_n^n - x_n^{-n} \\
\vdots & \vdots & \ddots & \vdots \\
x_1^2 - x_1^{-2} & x_2^2 - x_2^{-2} & \cdots & x_n^2 - x_n^{-2} \\
x_1^1 - x_1^{-1} & x_2^1 - x_2^{-1} & \cdots & x_n^1 - x_n^{-1}
\end{vmatrix}
$$

$$
= (x_1 x_2 \cdots x_n)^{-n} \prod_{j=1}^{n}(x_j^2 - 1) \prod_{1 \le i < j \le n}[(x_i - x_j)(x_i x_j - 1)]. \tag{6}
$$

Similar formulas can be proven by similar methods for the denominators of the characters of the orthogonal group, where there are two forms depending on whether the dimension is odd or even:

$$
\begin{vmatrix}
x_1^{n-1/2} - x_1^{-n+1/2} & x_2^{n-1/2} - x_2^{-n+1/2} & \cdots & x_n^{n-1/2} - x_n^{-n+1/2} \\
\vdots & \vdots & \ddots & \vdots \\
x_1^{3/2} - x_1^{-3/2} & x_2^{3/2} - x_2^{-3/2} & \cdots & x_n^{3/2} - x_n^{-3/2} \\
x_1^{1/2} - x_1^{-1/2} & x_2^{1/2} - x_2^{-1/2} & \cdots & x_n^{1/2} - x_n^{-1/2}
\end{vmatrix}
$$

$$
= (x_1 x_2 \cdots x_n)^{-n+1/2} \prod_{j=1}^{n}(x_j - 1) \prod_{1 \le i < j \le n}[(x_i - x_j)(x_i x_j - 1)], \tag{7}
$$

$$\begin{vmatrix} x_1^{n-1} + x_1^{-n+1} & x_2^{n-1} + x_2^{-n+1} & \cdots & x_n^{n-1} + x_n^{-n+1} \\ \vdots & \vdots & \ddots & \vdots \\ x_1^1 + x_1^{-1} & x_2^1 + x_2^{-1} & \cdots & x_n^1 + x_n^{-1} \\ x_1^0 + x_1^{-0} & x_2^0 + x_2^{-0} & \cdots & x_n^0 + x_n^{-0} \end{vmatrix}$$

$$= 2 \, (x_1 x_2 \cdots x_n)^{-n+1} \prod_{1 \le i < j \le n} [(x_i - x_j)(x_i x_j - 1)]. \tag{8}$$

3 A Different Generalization of the Vandermonde Product

There is a very different direction in which we can generalize the Vandermonde product. The starting point is the Desnanot–Jacobi adjoint matrix theorem [6, 8]. If we let M_j^i denote the minor of the matrix M obtained by removing row i and column j, then

$$\det M = \frac{\det M_1^1 \cdot \det M_n^n - \det M_1^n \cdot \det M_n^1}{\det M_{1,n}^{1,n}}. \tag{9}$$

If we define the determinant of the empty matrix to be 1 and the determinant of a 1×1 matrix to be the single entry in that matrix, then equation (9) provides a recursive definition of the general determinant. This recursive definition can be used to give a simple inductive proof of the Vandermonde determinant formula. If we modify the recursion, we get a modification of the Vandermonde determinant.

In 1980, David Robbins considered a λ-determinant defined recursively by

$$\det_\lambda M = \frac{\det_\lambda M_1^1 \cdot \det_\lambda M_n^n + \lambda \det_\lambda M_1^n \cdot \det_\lambda M_n^1}{\det_\lambda M_{1,n}^{1,n}}. \tag{10}$$

The usual determinant is the case $\lambda = -1$. If we apply this λ-determinant to the Vandermonde matrix $(x_j^{n-i})_{i,j=1}^n$, we get the product

$$\det_\lambda (x_j^{n-i}) = \prod_{1 \le i < j \le n} (x_i + \lambda x_j). \tag{11}$$

We would expect the λ-determinant of a general matrix (x_{ij}) to be extremely complicated. At each iteration we take a ratio of rational functions in λ and the x_i. What was surprising when Robbins first calculated these λ-determinants using *ALTRAN*, one of the early computer algebra systems, was that rather than intricate rational functions, the λ-determinants turned out to be predictable Laurent polynomials in λ and the x_j. For the 2×2 and 3×3 cases, this can be expected:

$$\det_\lambda \begin{pmatrix} a & b \\ c & d \end{pmatrix} = ad + \lambda bc,$$

$$\det_\lambda \begin{pmatrix} a\ b\ c \\ d\ e\ f \\ g\ h\ i \end{pmatrix} = aej + \lambda(bdj + afh) + \lambda^2(bfg + cdh) + \lambda^3 ceg$$

$$+\lambda(1 + \lambda)bde^{-1}fh.$$

The surprise came when it was discovered that all terms in the expansion of the λ-determinant of a matrix of any size were of the form a power of λ times a power of $1 + \lambda^{-1}$ times a monomial in x_{ij} and x_{ij}^{-1}. The monomials of an ordinary determinant can be expressed in terms of permutations. The monomials that Robbins found could all be expressed in terms of a generalization of the permutation matrix that he dubbed an *alternating-sign matrix*. This is a matrix such as

$$\begin{pmatrix} 0\ 1\ 0\ 0 \\ 1\ -1\ 0\ 1 \\ 0\ 0\ 0\ 1 \\ 0\ 1\ 0\ 0 \end{pmatrix},$$

where each row and each column sums to 1, and the nonzero entries in each row and column alternate in sign. Permutation matrices form a subset of the alternating-sign matrices.

Specifically, what Robbins and Rumsey [16] eventually proved was that

$$\det_\lambda(x_{i,j}) = \sum_{A=(a_{i,j})\in\mathcal{A}_n} \lambda^{\text{Inv}(A)}(1 + \lambda^{-1})^{N(A)} \prod_{i,j=1}^n x_{i,j}^{a_{i,j}}, \qquad (12)$$

where $\mathcal{A}_n$ is the set of $n \times n$ alternating-sign matrices, $N(A)$ is the number of -1's in A, and $\text{Inv}(A)$ is the inversion number of A, defined as the sum of $x_{i,j} \cdot x_{k,l}$ over all pairs $(x_{i,j}, x_{k,l})$ for which $i < k$ and $j > l$.

Thus, for example, the 4×4 alternating-sign matrix shown above corresponds to the term

$$\lambda^3(1 + \lambda^{-1})x_{1,2}x_{2,1}x_{2,2}^{-1}x_{3,4}x_{4,2}.$$

Another correspondence is given by

$$\begin{pmatrix} 0\ 1\ 0\ 0 \\ 1\ -1\ 1\ 0 \\ 0\ 1\ -1\ 1 \\ 0\ 0\ 1\ 0 \end{pmatrix} \longleftrightarrow \lambda^3(1 + \lambda^{-1})^2 x_{1,2}x_{2,1}x_{2,2}^{-1}x_{2,3}x_{3,2}x_{3,3}^{-1}x_{3,4}x_{4,3}.$$

An immediate corollary is that

$$\prod_{1\le i<j\le n}(x_i + \lambda x_j) = \sum_{A\in\mathcal{A}_n} \lambda^{\text{Inv}(A)}(1 + \lambda^{-1})^{N(A)} \prod_{i,j} x_j^{(n-i)a_{i,j}}. \qquad (13)$$

This is a remarkable formula. The formal expansion of the Vandermonde product creates $2^{n(n-1)/2}$ terms, but there is a great deal of cancellation, eliminating all but $n!$ terms. There is no cancellation when $\prod_{1\le i<j\le n}(x_i + \lambda x_j)$ is expanded into its $2^{n(n-1)/2}$ terms. Nevertheless, these terms can be collected as a sum over alternating-sign matrices. Similar λ-extensions of the other Weyl denominator formulas were published by Soichi Okada in 1993 [13].

4 Counting Alternating-Sign Matrices

An obvious question is to ask for the number of $n \times n$ alternating-sign matrices. How much smaller than $2^{n(n-1)/2}$ is it? Robbins and Rumsey found an efficient algorithm for counting these up to $n = 20$ and discovered that the cardinalities were products of suspiciously small primes:

$$|\mathcal{A}_1| = 1,$$
$$|\mathcal{A}_2| = 2,$$
$$|\mathcal{A}_3| = 7,$$
$$|\mathcal{A}_4| = 42 = 2 \cdot 3 \cdot 7,$$
$$|\mathcal{A}_5| = 429 = 3 \cdot 11 \cdot 13,$$
$$|\mathcal{A}_6| = 7436 = 2^2 \cdot 11 \cdot 13^2,$$
$$|\mathcal{A}_7| = 218348 = 2^2 \cdot 13^2 \cdot 17 \cdot 19,$$
$$|\mathcal{A}_8| = 10850216 = 2^3 \cdot \cdot 13 \cdot 17^2 \cdot 19^2,$$
$$|\mathcal{A}_9| = 911835460 = 2^2 \cdot 5 \cdot 17^2 \cdot 19^3 \cdot 23.$$

Robbins analyzed this sequence by taking each set $\mathcal{A}_n$ and splitting it into n disjoint subsets, where $\mathcal{A}_{n,k}$ consists of the $n \times n$ alternating-sign matrices with a 1 in row 1, column k. He created a Pascal-like triangle of these values, where $A_{n,k} = |\mathcal{A}_{n,k}|$ is the kth entry of row n:

$$
\begin{array}{ccccccccc}
 & & & & 1 & & & & \\
 & & & 1 & & 1 & & & \\
 & & 2 & & 3 & & 2 & & \\
 & 7 & & 14 & & 14 & & 7 & \\
42 & & 105 & & 135 & & 105 & & 42 \\
429 & 1287 & & 2002 & & 2002 & & 1287 & 429 \\
\end{array}
$$

We note that the first entry in each row is the sum of the entries in the row above, $A_{n,1} = A_{n-1}$. This is easy to justify, since $A_{n,1}$ counts alternating-sign matrices with a 1 in the upper left-hand corner, and the number of them is the number of ways of filling the $(n-1) \times (n-1)$ minor in the lower right-hand corner. A more profound observation is that the ratios of consecutive terms satisfy a curious pattern in which one adds numerators and denominators:

$$
\begin{array}{ccccccccccc}
 & & & & & 1 & & & & & \\
 & & & & 1 & 2/2 & 1 & & & & \\
 & & & 2 & 2/3 & 3 & 3/2 & 2 & & & \\
 & & 7 & 2/4 & 14 & 5/5 & 14 & 4/2 & 7 & & \\
 & 42 & 2/5 & 105 & 7/9 & 135 & 9/7 & 105 & 5/2 & 42 & \\
429 & 2/6 & 1287 & 9/14 & 2002 & 16/16 & 2002 & 14/9 & 1287 & 6/2 & 429 \\
\end{array}
$$

With a little effort, we can recognize the numerators and denominators as sums of binomial coefficients, leading to the first of the Mills, Robbins, Rumsey conjectures:

$$\frac{A_{n,k}}{A_{n,k+1}} = \frac{\binom{n-2}{k-1} + \binom{n-1}{k-1}}{\binom{n-2}{n-k-1} + \binom{n-1}{n-k-1}}. \tag{14}$$

These ratios together with the fact that $A_{n,1} = A_{n-1}$ and $A_1 = 1$ uniquely determine all values $A_{n,k}$. In particular, we get the second of the Mills, Robbins, Rumsey conjectures:

$$A_n = \prod_{j=0}^{n-1} \frac{(3j+1)!}{(n+j)!}. \tag{15}$$

Through Richard Stanley, Robbins learned that George Andrews had previously encountered this ratio of factorials and had proven that it counts D_n, the number of descending plane partitions that fit into an $n \times n \times n$ box. To prove the conjecture given in equation (15), they would only need to find a one-to-one correspondence between the plane partitions counted by D_n and the $n \times n$ alternating-sign matrices.

Though equations (14) and (15) have since been proven, a natural correspondence that would establish that these two sets are equinumerous is still unknown. The search for this correspondence, however, was almost immediately fruitful, suggesting questions for both plane partitions and alternating-sign matrices that would enable very rapid progress in both areas. This story is told in *Proofs and Confirmations* [3] (see also [4]).

5 More Conjectures

One important conjecture of Mills, Robbins, and Rumsey [12] concerns the polynomials defined by

$$A(n; x) = \sum_{A \in \mathcal{A}_n} x^{N(A)}.$$

A few specific values are given by

$A(1; x) = 1,$

$A(2; x) = 2,$

$A(3; x) = 6 + x,$

$A(4; x) = 24 + 16x + 2x^2,$

$A(5; x) = 120 + 200x + 94x^2 + 14x^3 + x^4,$

$A(6; x) = 720 + 2400x + 2684x^2 + 1284x^3 + 310x^4 + 36x^5 + 2x^6,$

$A(7; x) = 5040 + 24900x + 63308x^2 + 66158x^3 + 38390x^4 + 13037x^5 + 2660x^6$
$\qquad\qquad + 328x^7 + 26^8 + x^9.$

We see that $A(n; 0) = n!$, $A(n; 1) = A_n$, and (from equation (13)), $A(n; 2) = 2^{n(n-1)/2}$. Mills, Robbins, and Rumsey discovered that there is one more integer value of x that appears to yield a comparable formula:

$$A(n; 3) = \frac{3^{n(n-1)}}{2^{n(n-1)}} \prod_{\substack{1 \le i, j \le n \\ j-i \text{ odd}}} \frac{3(j - i) + 1}{3(j - i)}. \tag{16}$$

Another direction in which conjectures arose was from considering subsets of $\mathcal{A}_n$ that are invariant under certain simple transformations. For example, in 1991 Mills and Robbins [15] conjectured that

$$A_V(2n + 1) = (-3)^{n^2} \left(\prod_{\substack{1 \le i, j \le 2n+1 \\ 2 \mid j}} \frac{3(j - i) + 1}{j - i + 2n + 1} \right) A_n, \tag{17}$$

$$A_{HT}(2n) = (-3)^{n(n-1)/2} \left(\prod_{1 \le i, j \le n} \frac{3(j - i) + 2}{j - i + n} \right) A_n, \tag{18}$$

$$A_{QT}(4n) = A_{HT}(2n) \cdot A_n^2, \tag{19}$$

where $A_V(2n + 1)$ counts vertically symmetric $(2n + 1) \times (2n + 1)$ alternating-sign matrices, $A_{HT}(2n)$ counts $2n \times 2n$ alternating-sign matrices that are invariant under 180° rotation (half-turn), and $A_{QT}(4n)$ counts $4n \times 4n$ alternating-sign matrices that are invariant under 90° rotation (quarter-turn).

The first real breakthrough in the many conjectures surrounding alternating-sign matrices came with Doron Zeilberger's 1995 proof [19] of equation (15). It was an 84-page tour de force that was followed a year later by a much simpler proof by Greg Kuperberg [10], made possible by his recognition that there were results in statistical mechanics that were directly applicable to these questions. In particular, alternating-sign matrices are simply six-vertex models with a particular choice of boundary conditions.

6 The Six-Vertex Model

A six-vertex model is a directed graph on a square lattice as illustrated in Figure 3.1. Each vertex has in-degree and out-degree two, so that there are six possible configurations at each vertex. The boundary conditions that turn these into alternating-sign matrices are that there are in-edges along the left and right boundaries, out-edges along the top and bottom. The directed graph shown in the figure corresponds to the alternating-sign matrix

$$\begin{pmatrix} 0 & 1 & 0 & 0 & 0 \\ 1 & -1 & 0 & 1 & 0 \\ 0 & 1 & 0 & -1 & 1 \\ 0 & 0 & 0 & 1 & 0 \\ 0 & 0 & 1 & 0 & 0 \end{pmatrix}.$$

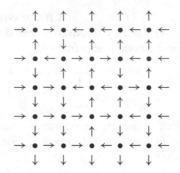

Fig. 3.1. Directed graph on a square lattice.

Horizontal vertices, $\rightarrow \bullet \leftarrow$ with $\uparrow$ above and $\downarrow$ below, correspond to $+1$, vertical vertices, $\leftarrow \bullet \rightarrow$ with $\downarrow$ above and $\uparrow$ below, corres-

pond to -1, and once these vertices have been placed, the orientations of all other directed edges are uniquely determined by the boundaries and the requirement that at each vertex, in-degree equals out-degree equals 2.

The number $N(A)$ of -1's in the alternating-sign matrix is the number of vertical vertices. The inversion number is equal to $N(A)$ plus the number of southwest

vertices, $\rightarrow \bullet \rightarrow$ with $\uparrow$ above and $\uparrow$ below.

To the vertex in row i, column j, we assign the pair of variables (x_i, y_j) and then define the weight of that vertex in the alternating-sign matrix or corresponding directed graph A by

$$
w_{i,j}(A) = \begin{cases} \sigma(a^2), & \text{if vertex is horizontal or vertical,} \\ \sigma(ax_i/y_j), & \text{if vertex is southeast or northwest,} \\ \sigma(ay_j/x_i), & \text{if vertex is southwest or northeast,} \end{cases}
$$

where $\sigma(z) = z - z^{-1}$.

The weight of A is the product of the weights of its vertices. The partition function sums the weight of A over all $A \in \mathcal{A}_n$. Anatoli Izergin [7] (see also [9]) recognized that this partition function could be expressed as a determinant:

$$
\sum_{A \in \mathcal{A}_n} \prod_{i,j} w_{i,j}(A)
$$

$$
= \det \left(\frac{1}{\sigma(ax_i/y_j) \cdot \sigma(ay_j/x_i)} \right) \frac{\prod_{i,j=1}^{n} \sigma(ax_i/y_j) \cdot \sigma(ay_j/x_i)}{\prod_{1 \le i < j \le n} \sigma(x_j/x_i) \cdot \sigma(y_i/y_j)} \sigma(a^2)^n. \quad (20)
$$

The proof is reminiscent of the proof of the Vandermonde determinant formula. If we multiply each side of equation (20) by the monomial $(x_1 \cdots x_n y_1 \cdots y_n)^{n-1}$, then each side becomes a polynomial in $x_1^2, \ldots, x_n^2, y_1^2, \ldots, y_n^2$. As a polynomial in x_1^2, each side has degree $n - 1$. The right side, an alternating function divided by Vandermonde products, is symmetric in the x_i^2 and symmetric in the y_j^2. We induct on n. It is fairly simple to show that if the formula is valid at $n - 1$, then the two expressions must be equal when $x_1 = y_1$. If the partition function on the left is also symmetric in the y_j^2, then these polynomials in x_1^2 agree at n values and so must be identical. The heart of the proof is showing that the partition function is a symmetric function in the x_i^2 and in the y_j^2. This follows from Rodney Baxter's triangle-to-triangle relation [1, 2].

If we now set $x_i = y_i = 1$ for all i, the partition function on the left side of equation (20) becomes

$$\sigma(a^2)^n \cdot \sigma(a)^{n^2-n} \sum_{A \in \mathcal{A}_n} (a + a^{-1})^{2N(A)}.$$

The right side requires some finesse, since simple substitution results in a highly singular matrix divided by a high power of 0. What Kuperberg did, in essence, was to set $x_i = q^i$, $y_j = q^j$, evaluate the determinant, and then take the limit $q \to 1$. For $a = e^{\pi i/3}$, we have $(a+a^{-1})^2 = 1$, and this yields the formula for A_n, equation (15). For $z = e^{\pi i/4}$, we get the formula for $A(n; 2)$. And for $z = e^{\pi i/6}$, this gives us a proof of the conjecture in equation (16).

Once the power of this determinant formula for the partition function and the triangle-to-triangle relation that lies behind it were recognized, more results followed. In 1996, Zeilberger [20] proved the original conjecture, equation (14). In 2001, Kuperberg [11] proved the three conjectures given in equations (17), (18), (19). He was able to prove much more. As an example, he defined UU alternating-sign matrices: one reads row $2i - 1$ from left to right and then comes back, reading right to left along row $2i$. The sum of the entries must be 1, and the nonzero entries must alternate in sign. At the same time, reading up column $2j - 1$ and then back down column $2j$, the sum of the entries must be 1 with nonzero entries alternating in sign. An example of such a UU alternating-sign matrix is

$$\begin{pmatrix} 1 & -1 & 0 & 1 \\ 0 & 1 & -1 & 0 \\ 0 & 0 & 0 & 0 \\ 0 & 0 & 1 & 0 \end{pmatrix}.$$

Kuperberg proved that the number of $2n \times 2n$ UU alternating-sign matrices is

$$A_{UU}(2n) = (-3)^{n^2} 2^{2n} \prod_{\substack{1 \le i, j \le 2n+1 \\ 2|j}} \frac{3(j-i)+2}{j-i+2n+1}. \tag{21}$$

In 2004, Okada [14] proved an important conjecture that had eluded Kuperberg, the product formula for the number of $(2n + 1) \times (2n + 1)$ alternating-sign matrices that are both vertically and horizontally symmetric. Significantly, Okada also tied these formulas back to the Weyl character formulas. The simplest and most striking of these is for the partition function for alternating-sign matrices with $a = e^{\pi i/3}$:

$$\sum_{A \in \mathcal{A}_n} \prod_{i,j} w_{i,j}(A)\Big|_{a=e^{\pi i/3}}$$

$$= 3^{-n(n-1)/2}(x_1 \cdots x_n y_1 \cdots y_n)^{1-n} s_\lambda(x_1^2, \ldots, x_n^2, y_1^2, \ldots, y_n^2), \qquad (22)$$

where s_λ is the Schur function indexed by the partition

$$\lambda = (n-1, n-1, n-2, n-2, \ldots, 1, 1, 0, 0).$$

In particular, the number of $n \times n$ alternating-sign matrices is $3^{-n(n-1)/2}$ times the dimension of the irreducible representation of GL_{2n} indexed by this λ. If we apply Schur's formula, equation (3), we see that

$$A_n = 3^{-n(n-1)/2} \prod_{1 \le i < j \le 2n} \frac{\lfloor (3j+1)/2 \rfloor - \lfloor (3i+1)/2 \rfloor}{j-i}, \qquad (23)$$

which, with a little arithmetic manipulation, can be recognized as $\prod_{j=0}^{n-1}(3j+1)!/(n+j)!$.

The partition function for vertically symmetric six-vertex models on a $(2n+1) \times (2n+1)$ lattice evaluated at $a = e^{\pi i/3}$ is $3^{-n(n-1)}$ times the symplectic character $\mathrm{SP}_{4n}(\lambda; x_1^2, \ldots, x_n^2, y_1^2, \ldots, y_n^2)$ defined in equation (5). The integer partition λ is the same, $\lambda = (n-1, n-1, \ldots, 0, 0)$. Okada proved the conjectured formulas for vertically and horizontally symmetric alternating-sign matrices by connecting the relevant partition functions to the characters of the symplectic group and of $\tilde{O}_{2n}$, the double cover of the orthogonal group O_{2n}.

There is more to do. Conjectured formulas for the number of half-turn symmetric, quarter-turn symmetric, and diagonally symmetric alternating-sign matrices have not been proven for odd dimensions. Okada has reformulated the half-turn and diagonal conjectures in terms of evaluations of Schur functions. More profoundly, we are only just beginning to understand what undergirds all of these relationships. Most of those who have worked on these problems recognize that these must be shadows of relationships that properly belong to representation theory. That is just another way of saying that, ultimately, this is really all about symmetry.

References

1. Rodney J. Baxter. Eight-vertex model in lattice statistics. *Physical Review Letters*, 26:832–833, 1971.
2. Rodney J. Baxter. *Exactly Solved Models in Statistical Mechanics*. Academic Press, London, 1982.
3. David M. Bressoud. *Proofs and Confirmations: The Story of the Alternating-Sign Matrix Conjecture*. Cambridge University Press, Cambridge, UK, 1999.

4. David M. Bressoud and James Propp. How the alternating-sign matrix conjecture was solved. *Notices Amer. Math. Soc.*, 46:637–646, 1999.

5. Augustin-Louis Cauchy. Mémoire sur les fonctions qui ne peuvent obtenir que deux valeurs égales et de signes contraires par suite des transpositions opérées entre les variables qu'elles renferment. *Journal de l'École Polytechnique*, 10(17):29–112, 1815. Reprinted in *Œuvres complètes d'Augustin Cauchy* series 2, Vol. 1, 91–161. Paris: Gauthier-Villars, 1899.

6. P. Desnanot. *Complément de la théorie des équations du premier degré*. Paris, 1819. Quoted in Thomas Muir, *The Theory of Determinants in the Historical Order of Development*, vol. 1, pp. 136–148. London: MacMillan and Co. 1906.

7. Anatoli G. Izergin. Partition function of a six-vertex model in a finite volume. *Dokl. Akad. Nauk SSSR*, 297:331–333, 1987.

8. C. G. J. Jacobi. De binis quibuslibet functionibus homogeneis secundi ordinis per substitutiones lineares in alias binas transformandis. *Journal für die Reine und Angewandte Mathematik*, 2:247–257, 1833. Reprinted in *C. G. J. Jacobi: Gesammelte Werke*. Vol. 3, pp. 191–268. Berlin: Georg Reimer, 1884.

9. Vladimir E. Korepin, Nikolai M. Bogoliubov, and Anatoli G. Izergin. *Quantum inverse scattering method and correlation functions*. Cambridge University Press, Cambridge, UK, 1993.

10. Greg Kuperberg. Another proof of the alternating-sign matrix conjecture. *International Mathematics Research Notes*, 1996:139–150, 1996.

11. Greg Kuperberg. Symmetry classes of alternating-sign matrices under one roof. *Ann. of Math.*, 156:835–866, 2002.

12. W. H. Mills, D. P. Robbins, and H. Rumsey. Alternating-sign matrices and descending plane partitions. *Journal of Combinatorial Theory*, 34:340–359, 1983.

13. Soichi Okada. Alternating-sign matrices and some deformations of Weyl's denominator formulas. *Journal of Algebraic Combinatorics*, 2:155–176, 1993.

14. Soichi Okada. Enumeration of symmetry classes of alternating-sign matrices and characters of classical groups. *Journal of Algebraic Combinatorics*, 23:43–69, 2006.

15. David P. Robbins. The story of 1, 2, 7, 42, 429, 7436, ... *The Mathematical Intelligencer*, 13:12–19, 1991.

16. David P. Robbins and Howard Rumsey. Determinants and alternating-sign matrices. *Advances in Mathematics*, 62:169–184, 1986.

17. I. J. Schur. Ein beitrag zur additiven Zahlentheorie und zur Theorie der Kettenbrüche. *S.-B. Preuss. Akad. Wiss. Phys.-Math. Kl.*, pages 302–321, 1917. Reprinted in *Gesammelte Abhandlungen*. Vol. 2, pp. 117–136. Berlin: Springer-Verlag, 1973.

18. Hermann Weyl. *The Classical Groups: Their Invariants and Representations*. Princeton University Press, Princeton, New Jersey, 1939.

19. Doron Zeilberger. Proof of the alternating-sign matrix conjecture. *Electronic Journal of Combinatorics*, 3, 1996. R13.

20. Doron Zeilberger. Proof of the refined alternating-sign matrix conjecture. *New York Journal of Mathematics*, 2:59–68, 1996.

4

Theta Functions in Complex Analysis and Number Theory

Hershel M. Farkas

Institute of Mathematics, The Hebrew University, Jerusalem 91904, Israel
farkas@math.huji.ac.il

Summary. In these notes we try to demonstrate the utility of the theory of theta functions in combinatorial number theory and complex analysis. The main idea is to use identities among theta functions to deduce either useful number-theoretic information related to representations as sums of squares and triangular numbers, statements concerning congruences, or statements concerning partitions of sets of integers. In complex analysis the main utility is in the theory of compact Riemann surfaces, with which we do not deal. We do show how identities among theta functions yield proofs of Picard's theorem and a conformal map of the rectangle onto the disk.

Key words: Theta functions, triangular numbers, partitions, Riemann map, Picard.

2000 *Mathematics Subject Classifications*: 11A, 11A25, 11B65, 30C

1 Introduction

This chapter is a slight elaboration of the lectures given by the author in Gainesville at the special year in 2005. The focus of the lectures given was an exposé of the use of theta functions in complex analysis and combinatorial number theory, illustrating how the theta function can be used in these areas. The theta function is absolutely essential in the theory of compact Riemann surfaces and abelian varieties. This is not quite true in the theory of functions of a complex variable and combinatorial number theory; however, these lectures were given with the purpose of showing their utility even here. Thus we begin with the definition of the n-dimensional theta function but will quite quickly restrict ourselves to the one-dimensional case.

We recall that if S is a compact Riemann surface of genus $g \geq 1$ and if $a_1, \ldots, a_g; b_1, \ldots, b_g$ is a canonical homology basis on S, then the dimension of the space of holomorphic differentials on S is g, and one can find a basis for this space, $(\theta_1, \ldots, \theta_g)$, with the property that

K. Alladi (ed.), *Surveys in Number Theory*, DOI: 10.1007/978-0-387-78510-3_4,
© Springer Science+Business Media, LLC 2008

$$\int_{a_j} \theta_i = \delta_{ij}, \int_{b_j} \theta_i = \tau_{ij},$$

where $\tau = \tau_{ij}$ is a symmetric matrix with positive definite imaginary part. The totality of $n \times n$ symmetric matrices with positive definite imaginary part is called the Siegel upper half-plane of degree n and will be denoted by H^n.

Definition 1. For $\zeta \in \mathbb{C}^n$ and $\tau \in H^n$,

$$\theta(\zeta, \tau) = \sum_{N \in \mathbb{Z}^n} \exp\left(2\pi i \left[\frac{1}{2} N \tau N + N\zeta\right]\right).$$

The function is quasiperiodic with respect to the group of translations

$$\langle \zeta \to \zeta + e^i, \zeta \to \zeta + \tau^i \rangle,$$

and it is easy to see that

$$\theta(\zeta + e^i, \tau) = \theta(\zeta, \tau),$$

$$\theta(\zeta + \tau^i, \tau) = \exp(l(\zeta))\theta(\zeta, \tau),$$

where $l(\zeta)$ is a linear function of ζ. More generally, one can make the following definition.

Definition 2. The theta function with characteristics

$$\begin{bmatrix} \epsilon \\ \epsilon' \end{bmatrix}$$

is defined by

$$\theta\begin{bmatrix} \epsilon \\ \epsilon' \end{bmatrix}(\zeta, \tau)$$

$$= \sum_{N \in \mathbb{Z}^n} \exp\left(2\pi i \left[\frac{1}{2}\left(N + \frac{\epsilon}{2}\right)\tau\left(N + \frac{\epsilon}{2}\right) + \left(N + \frac{\epsilon}{2}\right)\left(\zeta + \frac{\epsilon'}{2}\right)\right]\right).$$

It is easy to see that once again we have

$$\theta\begin{bmatrix} \epsilon \\ \epsilon' \end{bmatrix}(\zeta + \tau N + IM, \tau) = \exp(l(\zeta))\theta\begin{bmatrix} \epsilon \\ \epsilon' \end{bmatrix}(\zeta, \tau).$$

For applications of this theory to compact Riemann surfaces the reader is referred to the monographs [6], [7], and [4] as well as the references contained therein, especially the many articles of R. Accola [9].

2 ζ-Theory for $n = 1$

We now consider the case $n = 1$, and will write some exact formulas.

$$\theta \begin{bmatrix} \epsilon \\ \epsilon' \end{bmatrix} (\zeta + n + m\tau, \tau) = \exp(2\pi i) \left[\frac{n\epsilon - m\epsilon'}{2} - m\zeta - \frac{m^2\tau}{2} \right] \theta \begin{bmatrix} \epsilon \\ \epsilon' \end{bmatrix} (\zeta, \tau),$$

(1)

$$\theta \begin{bmatrix} \epsilon + 2m \\ \epsilon' + 2n \end{bmatrix} (\zeta, \tau) = \exp(\pi i \epsilon n) \theta \begin{bmatrix} \epsilon \\ \epsilon' \end{bmatrix} (\zeta, \tau).$$

(2)

For m, n real numbers,

$$\theta \begin{bmatrix} \epsilon \\ \epsilon' \end{bmatrix} \left(\zeta + \frac{n + m\tau}{2}, \tau \right)$$

$$= \exp(2\pi i) \left[\frac{-m\zeta}{2} - \frac{m^2\tau}{8} - \frac{m(\epsilon' + n)}{4} \right] \theta \begin{bmatrix} \epsilon + m \\ \epsilon' + n \end{bmatrix} (\zeta, \tau),$$

(3)

$$\theta \begin{bmatrix} \epsilon \\ \epsilon' \end{bmatrix} (\zeta, \tau) = \sum_{l=0}^{N-1} \theta \begin{bmatrix} \frac{\epsilon + 2l}{N} \\ N\epsilon' \end{bmatrix} (N\zeta, N^2\tau),$$

(4)

$$\frac{\partial^2}{\partial \zeta^2} \theta \begin{bmatrix} \epsilon \\ \epsilon' \end{bmatrix} (\zeta, \tau) = 4\pi i \frac{\partial}{\partial \tau} \theta \begin{bmatrix} \epsilon \\ \epsilon' \end{bmatrix} (\zeta, \tau).$$

(5)

It is a fascinating fact of life that the theory is already quite rich when

$$\begin{bmatrix} \epsilon \\ \epsilon' \end{bmatrix} \in \mathbb{Z}^2.$$

Making use of equation (2), we see that there are only four theta functions that need concern us here:

$$\theta \begin{bmatrix} 0 \\ 0 \end{bmatrix} (\zeta, \tau), \quad \theta \begin{bmatrix} 0 \\ 1 \end{bmatrix} (\zeta, \tau), \quad \theta \begin{bmatrix} 1 \\ 0 \end{bmatrix} (\zeta, \tau), \quad \theta \begin{bmatrix} 1 \\ 1 \end{bmatrix} (\zeta, \tau).$$

It is rather easy to see that

$$\theta \begin{bmatrix} -\epsilon \\ -\epsilon' \end{bmatrix} (\zeta, \tau) = \theta \begin{bmatrix} \epsilon \\ \epsilon' \end{bmatrix} (-\zeta, \tau).$$

Thus from equation (2) above,

$$\theta \begin{bmatrix} \epsilon \\ \epsilon' \end{bmatrix} (-\zeta, \tau) = \theta \begin{bmatrix} -\epsilon \\ -\epsilon' \end{bmatrix} (\zeta, \tau)$$

$$= \theta \begin{bmatrix} \epsilon - 2\epsilon \\ \epsilon' - 2\epsilon' \end{bmatrix} (\zeta, \tau) = \exp(-\pi i \epsilon \epsilon') \theta \begin{bmatrix} \epsilon \\ \epsilon' \end{bmatrix} (\zeta, \tau).$$

Hence the first three functions are even, and the last one is odd. The zeros of these functions are well understood. There is only one zero in the fundamental parallelogram, and the zero of $\theta\left[\begin{smallmatrix}\epsilon\\\epsilon'\end{smallmatrix}\right](\zeta, \tau)$ is at the point

$$\frac{1-\epsilon}{2}\tau + \frac{1-\epsilon'}{2}.$$

All the above has been what I shall call the ζ-theory.

3 τ-Theory

The main theorem of the τ-theory, as opposed to the ζ-theory, is the following:

Theorem 1. *For any $\left[\begin{smallmatrix}\epsilon\\\epsilon'\end{smallmatrix}\right] \in R^2$ and any $\gamma \in \mathrm{SL}(2, \mathbb{Z})$ with $\gamma = \left[\begin{smallmatrix}a & b\\c & d\end{smallmatrix}\right]$, we have*

$$\exp\left(-\pi i \frac{c\zeta^2}{c\tau+d}\right) \theta\left[\begin{matrix}\epsilon\\\epsilon'\end{matrix}\right]\left(\frac{\zeta}{c\tau+d}, \gamma(\tau)\right)$$

$$= \kappa\left(\left[\begin{matrix}\epsilon\\\epsilon'\end{matrix}\right], \gamma\right)\sqrt{c\tau+d}\,\theta\left[\begin{matrix}a\epsilon+c\epsilon'-ac\\b\epsilon+d\epsilon'+bd\end{matrix}\right](\zeta, \tau).$$

In particular, when $\zeta = 0$ we have

$$\theta\left[\begin{matrix}\epsilon\\\epsilon'\end{matrix}\right](0, \gamma(\tau))\gamma'(\tau)^{\frac{1}{4}} = \kappa\left(\left[\begin{matrix}\epsilon\\\epsilon'\end{matrix}\right], \gamma\right)\sqrt{c\tau+d}\,\theta\left[\begin{matrix}a\epsilon+c\epsilon'-ac\\b\epsilon+d\epsilon'+bd\end{matrix}\right](0, \tau)$$

and

$$\kappa\left(\left[\begin{matrix}\epsilon\\\epsilon'\end{matrix}\right], \gamma\right)$$

$$= \exp(2\pi i)\left[\frac{1}{4}(a\epsilon+c\epsilon')bd - \frac{1}{8}(ab\epsilon^2 + cd\epsilon'^2 + 2bc\epsilon\epsilon')\right] \cdot \kappa\left(\left[\begin{matrix}0\\0\end{matrix}\right], \gamma\right).$$

Finally, we have also

$$\kappa^8\left(\left[\begin{matrix}0\\0\end{matrix}\right], \gamma\right) = 1.$$

We will content ourselves with just this theorem. We will not prove this result here, but the interested reader can find the proof in [5].

4 Applications of the Above Ideas

There is an additional property of theta functions that we shall need in the sequel but postpone it to the next section. Here, we prove a theta function identity that we will show ultimately is the basis of the construction of a conformal map of the rectangle onto the unit disk as well as the basic result that allows us to prove Jacobi's quantitative version of Lagrange's theorem that each positive integer is representable as a sum of four squares.

Theorem 2.

$$\left(\frac{\theta'\begin{bmatrix} 1 \\ 1 \end{bmatrix}}{\theta\begin{bmatrix} 0 \\ 0 \end{bmatrix}\theta\begin{bmatrix} 0 \\ 1 \end{bmatrix}\theta\begin{bmatrix} 1 \\ 0 \end{bmatrix}} \right)^2 = \frac{\theta''\begin{bmatrix} 0 \\ 1 \end{bmatrix}}{\theta\begin{bmatrix} 0 \\ 1 \end{bmatrix}} - \frac{\theta''\begin{bmatrix} 1 \\ 0 \end{bmatrix}}{\theta\begin{bmatrix} 1 \\ 0 \end{bmatrix}}$$

$$= \frac{\theta''\begin{bmatrix} 0 \\ 0 \end{bmatrix}}{\theta\begin{bmatrix} 0 \\ 0 \end{bmatrix}} - \frac{\theta''\begin{bmatrix} 1 \\ 0 \end{bmatrix}}{\theta\begin{bmatrix} 1 \\ 0 \end{bmatrix}} = \frac{\theta''\begin{bmatrix} 0 \\ 1 \end{bmatrix}}{\theta\begin{bmatrix} 0 \\ 1 \end{bmatrix}} - \frac{\theta''\begin{bmatrix} 0 \\ 0 \end{bmatrix}}{\theta\begin{bmatrix} 0 \\ 0 \end{bmatrix}}.$$

Proof. The proof is totally elementary and relies on the well-known fact that an elliptic function cannot have as its only singularity in the fundamental parallelogram a simple pole. We begin with

$$f(\zeta) = \frac{\theta^2\begin{bmatrix} 0 \\ 1 \end{bmatrix}(\zeta,\tau)}{\theta^2\begin{bmatrix} 0 \\ 0 \end{bmatrix}(\zeta,\tau)}, \quad g(\zeta) = \frac{\theta^2\begin{bmatrix} 1 \\ 1 \end{bmatrix}(\zeta,\tau)}{\theta^2\begin{bmatrix} 0 \\ 0 \end{bmatrix}(\zeta,\tau)},$$

and observe that they are elliptic functions with periods $1, \tau$ and have double poles at $\frac{\tau+1}{2}$. This easily implies that the three functions $1, f, g$ are linearly dependent, and we easily compute

$$\theta^2\begin{bmatrix} 0 \\ 0 \end{bmatrix}\theta^2\begin{bmatrix} 0 \\ 1 \end{bmatrix}(\zeta,\tau) - \theta^2\begin{bmatrix} 1 \\ 0 \end{bmatrix}\theta^2\begin{bmatrix} 1 \\ 1 \end{bmatrix}(\zeta,\tau) - \theta^2\begin{bmatrix} 0 \\ 1 \end{bmatrix}\theta^2\begin{bmatrix} 0 \\ 0 \end{bmatrix}(\zeta,\tau) = 0.$$

$$(6)$$

Expanding this identity in a Taylor series about the origin and focusing on the coefficient of z^2 gives the first identity of the theorem.

This identity gives rise to others by, for example, setting

$$\zeta = z + \frac{1}{2}.$$

We get

$$\theta^2\begin{bmatrix} 0 \\ 0 \end{bmatrix}\theta^2\begin{bmatrix} 0 \\ 0 \end{bmatrix}(z,\tau) - \theta^2\begin{bmatrix} 1 \\ 0 \end{bmatrix}\theta^2\begin{bmatrix} 1 \\ 0 \end{bmatrix}(z,\tau) - \theta^2\begin{bmatrix} 0 \\ 1 \end{bmatrix}\theta^2\begin{bmatrix} 0 \\ 1 \end{bmatrix}(z,\tau) = 0. \quad (7)$$

$\square$

5 Infinite Products

In this section we give the last property of theta functions that we want to mention. The property is that the theta functions all have infinite product expansions. This is of course a dimension-1 phenomenon that does not persist in higher dimensions. Everything else that we have done till now has a multidimensional analogue.

We make the change of variables $x = \exp(\pi i \tau)$, $z = \exp(2\pi i \zeta)$. Jacobi then showed that

$$\theta \begin{bmatrix} \epsilon \\ \epsilon' \end{bmatrix} (\zeta, \tau) = \exp\left(\frac{\pi i \epsilon \epsilon'}{2}\right) x^{\frac{\epsilon^2}{4}} z^{\frac{\epsilon}{2}}$$

$$\cdot \prod_{n=1}^{\infty} (1 - x^{2n})(1 + e^{\pi i \epsilon'} x^{2n-1+\epsilon} z)(1 + e^{-\pi i \epsilon'} x^{2n-1-\epsilon}/z).$$

Thus we have

$$\theta \begin{bmatrix} 0 \\ 0 \end{bmatrix} (\zeta, \tau) = \prod_{n=1}^{\infty} (1 - x^{2n})(1 + x^{2n-1}z)(1 + x^{2n-1}/z),$$

$$\theta \begin{bmatrix} 0 \\ 1 \end{bmatrix} (\zeta, \tau) = \prod_{n=1}^{\infty} (1 - x^{2n})(1 - x^{2n-1}z)(1 - x^{2n-1}/z),$$

$$\theta \begin{bmatrix} 1 \\ 0 \end{bmatrix} (\zeta, \tau) = x^{\frac{1}{4}} z^{\frac{1}{2}} \prod_{n=1}^{\infty} (1 - x^{2n})(1 + x^{2n}z)(1 + x^{2n-2}/z).$$

We shall see that it is this property that will be the source of the combinatorial number theory that we discuss.

6 Combinatorial Number Theory

In this section we shall introduce some problems in combinatorial number theory. Our focus at first will be the well-known result that every positive integer is the sum of four squares. We will give a complete proof of this result using only what we have discussed till now. In fact, we shall prove Jacobi's result, which says that the number of representations of the positive integer N as a sum of four squares is given by eight times the sum of the divisors of N that are not congruent to 0 mod 4. We begin with the observation that in the variable $x = \exp(\pi i \tau)$,

$$\theta \begin{bmatrix} 0 \\ 0 \end{bmatrix} (0, \tau) = \sum_{n=-\infty}^{\infty} x^{n^2},$$

$$\theta \begin{bmatrix} 0 \\ 1 \end{bmatrix} (0, \tau) = \sum_{n=-\infty}^{\infty} (-1)^n x^{n^2},$$

and

$$\theta \begin{bmatrix} 1 \\ 0 \end{bmatrix} (0, \tau) = x^{\frac{1}{4}} \sum_{n=-\infty}^{\infty} x^{n(n+1)} = x^{\frac{1}{4}} \sum_{n=-\infty}^{\infty} x^{\frac{2n(n+1)}{2}}.$$

Definition 3. $S_k(N)$ is the number of integer solutions to the equation

$$x_1^2 + \cdots + x_k^2 = N;$$

$T_k(N)$ is the number of integer solutions to the equation

$$\frac{x_1(x_1 + 1)}{2} + \cdots + \frac{x_k(x_k + 1)}{2} = N.$$

Setting $z = 0$ in equation (7) above gives

$$\theta^4 \begin{bmatrix} 0 \\ 0 \end{bmatrix} (0, \tau) - \theta^4 \begin{bmatrix} 0 \\ 1 \end{bmatrix} (0, \tau) - \theta^4 \begin{bmatrix} 1 \\ 0 \end{bmatrix} (0, \tau) = 0.$$

In the variable $x = e^{\pi i \tau}$, this identity reads

$$\sum_{n=1}^{\infty} S_4(n)(1 - (-1)^n)x^n = \sum_{n=0}^{\infty} T_4(n)x^{2n+1},$$

or clearly

$$\sum_{n=0}^{\infty} 2S_4(2n + 1)x^{2n+1} = \sum_{n=0}^{\infty} T_4(n)x^{2n+1}.$$

It thus follows that for all nonnegative integers n,

$$2S_4(2n + 1) = T_4(n).$$

What, however, is $S_4(n)$? This is the question that Jacobi asked and answered. The answer is already found in the first identity of Theorem 2. In Theorem 2 we showed that

$$\left(\frac{\theta' \begin{bmatrix} 1 \\ 1 \end{bmatrix} (0, \tau)}{\theta \begin{bmatrix} 0 \\ 0 \end{bmatrix} (0, \tau)\theta \begin{bmatrix} 0 \\ 1 \end{bmatrix} (0, \tau)\theta \begin{bmatrix} 1 \\ 0 \end{bmatrix} (0, \tau)} \right)^2 = \frac{\dfrac{\theta'' \begin{bmatrix} 0 \\ 1 \end{bmatrix} (0,\tau)}{\theta \begin{bmatrix} 0 \\ 1 \end{bmatrix} (0,\tau)} - \dfrac{\theta'' \begin{bmatrix} 1 \\ 0 \end{bmatrix} (0,\tau)}{\theta \begin{bmatrix} 1 \\ 0 \end{bmatrix} (0,\tau)}}{\theta^4 \begin{bmatrix} 0 \\ 0 \end{bmatrix} (0, \tau)}.$$

The first point is to see that the left-hand side is actually constant in the variable τ. This follows from the transformation formula. If we raise the left-hand side to the fourth power, it is easily seen to be automorphic with respect to the group PSL(2, $\mathbb{Z}$). Since $H/\text{PSL}(2, \mathbb{Z})$ is a punctured sphere, the complex plane, and since the function is bounded as one goes to infinity, it must be constant. Checking, we find the value to be π^8. Taking now the fourth root, we have the left-hand side equal to π^2.

Equation (5) above allows us to write the right-hand side as

$$4\pi i \left(\frac{\partial}{\partial \tau} \log \theta \begin{bmatrix} 0 \\ 1 \end{bmatrix} (0, \tau) - \frac{\partial}{\partial \tau} \log \theta \begin{bmatrix} 1 \\ 0 \end{bmatrix} (0, \tau) \right).$$

Using the product representations of these functions and the fact that

$$\frac{\partial}{\partial \tau} = \pi i x \frac{\partial}{\partial x},$$

we get

$$-4\pi^2 x \left(\sum_{n=1}^{\infty} \left[\frac{-2(2n-1)x^{2n-2}}{1 - x^{2n-1}} + \frac{-2 \cdot 2n x^{2n-1}}{1 + x^{2n}} \right] - \frac{1}{4x} \right).$$

It thus follows that

$$\theta^4 \begin{bmatrix} 0 \\ 0 \end{bmatrix} = 1 + 8 \left(\sum_{n=1}^{\infty} \frac{(2n-1)x^{2n-1}}{1 - x^{2n-1}} + \frac{2n x^{2n}}{1 + x^{2n}} \right).$$

This last expression can be rewritten

$$1 + 8 \left(\sum_{n=1}^{\infty} \frac{(2n-1)x^{2n-1}}{1 - x^{2n-1}} + \frac{2n x^{2n}}{1 - x^{2n}} - \frac{2n x^{2n}}{1 - x^{2n}} + \frac{2n x^{2n}}{1 + x^{2n}} \right),$$

which can be replaced by

$$1 + 8 \left(\sum_{n=1}^{\infty} \frac{n x^n}{1 - x^n} - \frac{4n x^{4n}}{1 - x^{4n}} \right),$$

which we see is

$$1 + 8 \left(\sum_{n=1}^{\infty} \left[\sigma(n) - 4\sigma \left(\frac{n}{4} \right) \right] \right) x^n,$$

which equals

$$1 + 8 \sum_{n=1}^{\infty} \tilde{\sigma}(n) x^n,$$

and $\tilde{\sigma}(n)$ is the sum of the divisors of n not congruent to 0 mod 4. We have just proven the following.

Theorem 3. *The number of representations of the positive integer N as a sum of four squares is given by eight times the sum of the divisors of N not congruent to zero mod 4.*

7 Sums of Squares and Triangular Numbers

In the previous section we derived Jacobi's result that

$$S_4(n) = 8\tilde{\sigma}(n).$$

A simpler result of Jacobi is

$$S_2(n) = 4(d_1(n) - d_3(n)),$$

where $d_i(n)$ is the number of divisors of n congruent to i mod 4. From this formula we can conclude

$$S_2(4n + 3) = 0, \ S_2(2n) = S_2(n).$$

These results do not require Jacobi's results but follow immediately from them. We shall show how these results also follow from some simple theta constant identities and then generalize a bit. The results in this section are contained in [3]. We need the following identity and some of its consequences:

$$\theta \begin{bmatrix} \epsilon \\ \epsilon' \end{bmatrix} (0, \tau) \theta \begin{bmatrix} \delta \\ \delta' \end{bmatrix} (0, \tau) = \theta \begin{bmatrix} \frac{\epsilon+\delta}{2} \\ \epsilon' + \delta' \end{bmatrix} (0, 2\tau) \theta \begin{bmatrix} \frac{\epsilon-\delta}{2} \\ \epsilon' - \delta' \end{bmatrix} (0, 2\tau)$$

$$+ \theta \begin{bmatrix} \frac{\epsilon+\delta}{2} + 1 \\ \epsilon' + \delta' \end{bmatrix} (0, 2\tau) \theta \begin{bmatrix} \frac{\epsilon-\delta}{2} + 1 \\ \epsilon' - \delta' \end{bmatrix} (0, 2\tau), \quad (8)$$

$$\theta^2 \begin{bmatrix} 0 \\ 0 \end{bmatrix} (0, \tau) = \theta^2 \begin{bmatrix} 0 \\ 0 \end{bmatrix} (0, 2\tau) + \theta^2 \begin{bmatrix} 1 \\ 0 \end{bmatrix} (0, 2\tau), \quad (9)$$

$$\theta^2 \begin{bmatrix} 0 \\ 1 \end{bmatrix} (0, \tau) = \theta^2 \begin{bmatrix} 0 \\ 0 \end{bmatrix} (0, 2\tau) - \theta^2 \begin{bmatrix} 1 \\ 0 \end{bmatrix} (0, 2\tau), \quad (10)$$

$$\theta^2 \begin{bmatrix} 1 \\ 0 \end{bmatrix} (0, \tau) = 2\theta \begin{bmatrix} 0 \\ 0 \end{bmatrix} (0, 2\tau) \theta \begin{bmatrix} 1 \\ 0 \end{bmatrix} (0, 2\tau). \quad (11)$$

In the variable $x = \exp(\pi i \tau)$, equation (9) above says that

$$\sum_{n=1}^{\infty} S_2(n) x^n = \sum_{n=1}^{\infty} S_2(n) x^{2n} + \sum_{n=0}^{\infty} T_2(n) x^{4n+1},$$

from which we immediately conclude

$$S_2(4n + 3) = 0, \ S_2(4n + 1) = T_2(n), \ S_2(2n) = S_2(n).$$

Return now to equation (4) and observe that we get the following instances of it corresponding to $N = 2, 3$:

$$\theta \begin{bmatrix} 0 \\ 0 \end{bmatrix} (0, \tau) = \theta \begin{bmatrix} 0 \\ 0 \end{bmatrix} (0, 4\tau) + \theta \begin{bmatrix} 1 \\ 0 \end{bmatrix} (0, 4\tau), \quad (12)$$

$$\theta\begin{bmatrix} 0 \\ 1 \end{bmatrix}(0,\tau) = \theta\begin{bmatrix} 0 \\ 0 \end{bmatrix}(0,4\tau) - \theta\begin{bmatrix} 1 \\ 0 \end{bmatrix}(0,4\tau),\tag{13}$$

$$\theta\begin{bmatrix} 1 \\ 0 \end{bmatrix}(0,\tau) = 2\theta\begin{bmatrix} \frac{1}{3} \\ 0 \end{bmatrix}(0,9\tau) + \theta\begin{bmatrix} 1 \\ 0 \end{bmatrix}(0,9\tau),\tag{14}$$

$$\theta\begin{bmatrix} 0 \\ 0 \end{bmatrix}(0,\tau) = 2\theta\begin{bmatrix} \frac{2}{3} \\ 0 \end{bmatrix}(0,9\tau) + \theta\begin{bmatrix} 0 \\ 0 \end{bmatrix}(0,9\tau).\tag{15}$$

In the last two equations I have used the elementary facts that

$$\theta\begin{bmatrix} \frac{1}{3} \\ 0 \end{bmatrix}(0,9\tau) = \theta\begin{bmatrix} \frac{5}{3} \\ 0 \end{bmatrix}(0,9\tau),$$

$$\theta\begin{bmatrix} \frac{4}{3} \\ 0 \end{bmatrix}(0,9\tau) = \theta\begin{bmatrix} \frac{2}{3} \\ 0 \end{bmatrix}(0,9\tau).$$

Cubing equations (12) and (13), adding, and subtracting, we get the following result.

Lemma 1. *For all $\tau \in H$ we have*

$$\theta^3\begin{bmatrix} 0 \\ 0 \end{bmatrix}(0,\tau) + \theta^3\begin{bmatrix} 0 \\ 1 \end{bmatrix}(0,\tau) = 2\theta^3\begin{bmatrix} 0 \\ 0 \end{bmatrix}(0,4\tau) + 6\theta\begin{bmatrix} 0 \\ 0 \end{bmatrix}(0,4\tau)\theta^2\begin{bmatrix} 1 \\ 0 \end{bmatrix}(0,4\tau),\tag{16}$$

$$\theta^3\begin{bmatrix} 0 \\ 0 \end{bmatrix}(0,\tau) - \theta^3\begin{bmatrix} 0 \\ 1 \end{bmatrix}(0,\tau) = 2\theta^3\begin{bmatrix} 1 \\ 0 \end{bmatrix}(0,4\tau) + 6\theta^2\begin{bmatrix} 0 \\ 0 \end{bmatrix}(0,4\tau)\theta\begin{bmatrix} 1 \\ 0 \end{bmatrix}(0,4\tau).\tag{17}$$

In the variable $x = \exp(\pi i\tau)$, equations (16), (17) become

$$2\sum_{n=0}^{\infty} S_3(2n)x^{2n} = 2\sum_{n=0}^{\infty} S_3(n)x^{4n} + 6x^2\sum_{n=-\infty}^{\infty} x^{4n^2}\sum_{n=0}^{\infty} T_2(n)x^{8n},$$

$$2\sum_{n=0}^{\infty} S_3(2n+1)x^{2n+1} = 2x^3\sum_{n=0}^{\infty} T_3(n)x^{8n} + 6x\sum_{n=-\infty}^{\infty} x^{8\frac{n^2+n}{2}}\sum_{n=0}^{\infty} S_2(n)x^{4n}.$$

We have the following immediate corollary.

Corollary 1.

$$S_3(4k) = S_3(k), \; S_3(8k+7) = 0, \; S_3(8k+3) = T_3(k).$$

The above facts about $S_3(n)$, especially the middle statement, are not difficult to prove, but here they come without any pain. The last statement is the basis of Gauss's proof that every nonnegative integer is the sum of three triangular numbers.

Let us now turn to equation (14). In the variable $x = \exp(\pi i\tau)$ we find after cubing and replacing x by $x^{\frac{1}{2}}$ that

$$\left(\sum_{n=-\infty}^{\infty} x^{\frac{n^2+n}{2}}\right)^3 = 8\left(\sum_{n=-\infty}^{\infty} x^{3\frac{3n^2+n}{2}}\right)^3 + 12x\left(\sum_{n=-\infty}^{\infty} x^{3\frac{3n^2+n}{2}}\right)^2$$

$$\times \sum_{n=-\infty}^{\infty} x^{9\frac{n^2+n}{2}} + 6x^2 \sum_{n=-\infty}^{\infty} x^{3\frac{3n^2+n}{2}}$$

$$\times \left(\sum_{n=-\infty}^{\infty} x^{9\frac{n^2+n}{2}}\right)^2 + x^3 \left(\sum_{n=-\infty}^{\infty} x^{9\frac{n^2+n}{2}}\right)^3.$$

Definition 4. Let $P_k(n)$ denote the number of solutions of the Diophantine equation

$$\frac{3x_1^2 + x_1}{2} + \cdots + \frac{3x_k^2 + x_k}{2} = n.$$

Clearly $P_k(n)$ is the number of ways n can be written as a sum of k generalized pentagonal numbers. We recall that a pentagonal number is a number of the form $\frac{3k^2-k}{2}$ with k nonnegative. Our previous equation can now be rewritten as

$$\sum_{n=0}^{\infty} T_3(n)x^n = 8 \sum_{n=0}^{\infty} P_3(n)x^{3n} + 12x \sum_{n=0}^{\infty} P_2(n)x^{3n} \sum_{n=-\infty}^{\infty} x^{9\frac{n^2+n}{2}}$$

$$+ 6x^2 \sum_{n=0}^{\infty} T_2(n)x^{9n} \sum_{n=-\infty}^{\infty} x^{3\frac{3n^2+n}{2}} + x^3 \sum_{n=0}^{\infty} T_3(n)x^{9n}.$$

An immediate consequence is the following:

Corollary 2. *For all positive integers k we have*

$$T_3(3k) - 8P_3(k) \mid T_3\left(\frac{k-1}{3}\right),$$

where T_k is defined to be 0 whenever the variable is not a nonnegative integer. Hence $T_3(k) = 8P_3(k)$ unless $k \equiv 1 \bmod 3$:

$$T_3(3k+2) = 6\sum_{l\in\mathbb{Z}} T_2\left(\frac{k - \frac{3l^2+l}{2}}{3}\right).$$

In particular, $T_3(3k+2)$ is congruent to $0 \bmod 24$.

All of the preceding follows from our identity above.

Consider now equation (12). Cubing it leads to

$$\theta^3\begin{bmatrix}0\\0\end{bmatrix}(0,\tau) = \theta^3\begin{bmatrix}0\\0\end{bmatrix}(0,4\tau) + 3\theta^2\begin{bmatrix}0\\0\end{bmatrix}(0,4\tau)\theta\begin{bmatrix}1\\0\end{bmatrix}(0,4\tau)$$

$$+ 3\theta\begin{bmatrix}0\\0\end{bmatrix}(0,4\tau)\theta^2\begin{bmatrix}1\\0\end{bmatrix}(0,4\tau) + \theta^3\begin{bmatrix}1\\0\end{bmatrix}(0,4\tau).$$

Using equation (12) again followed by equation (11) gives

$$\theta^3 \begin{bmatrix} 0 \\ 0 \end{bmatrix}(0,\tau) = \theta^3 \begin{bmatrix} 0 \\ 0 \end{bmatrix}(0,4\tau) + \theta^3 \begin{bmatrix} 1 \\ 0 \end{bmatrix}(0,4\tau) + \frac{3}{2}\theta \begin{bmatrix} 0 \\ 0 \end{bmatrix}(0,\tau)\theta^2 \begin{bmatrix} 1 \\ 0 \end{bmatrix}(0,2\tau).$$

(18)

Equation (18) is the main identity we shall use from here on. In the variable $x = \exp(\pi i \tau)$ it reads

$$\sum_{n=0}^{\infty} S_3(n)x^n = \sum_{n=0}^{\infty} S_3(n)x^{4n} + x^3 \sum_{n=0}^{\infty} T_3(n)x^{8n} + \frac{3}{2}x \sum_{n=-\infty}^{\infty} x^{n^2} \sum_{n=0}^{\infty} T_2(n)x^{4n}.$$

The above equation implies some old information, namely

$$S_3(8N+3) = T_3(N), \quad S_3(4N) = S_3(N), \quad S_3(8N+7) = 0.$$

There is, however, some new information embedded in the term

$$\frac{3}{2}x \sum_{n=-\infty}^{\infty} x^{n^2} \sum_{n=0}^{\infty} T_2(n)x^n.$$

We rewrite this as a sum of two terms:

$$\frac{3}{2}x \sum_{n=0}^{\infty} 2x^{(2n)^2} \sum_{n=0}^{\infty} T_2\left(\frac{n}{4}\right)x^n + \frac{3}{2}x \sum_{n=0}^{\infty} 2x^{(2n+1)^2} \sum_{n=0}^{\infty} T_2\left(\frac{n}{4}\right)x^n.$$

The first term contains all exponents congruent to 1 mod 4, and the second term contains all exponents congruent to 2 mod 4.

Theorem 4.

$$S_3(4N+1) = \frac{3}{2}\left(\sum_{k=1}^{\infty} 2T_2(N-k^2) + T_2(N)\right),$$

$$S_3(4N+2) = 3\sum_{k=0}^{\infty} T_2(N-k^2-k).$$

Proof. Compute the Cauchy product of the power series.   □

Corollary 3. *Every positive integer is the sum of two triangular numbers and a square.*

Proof. Since $S_3(4N+1)$ is positive for every N, for at least one k it must be the case that $T_2(N-k^2)$ is positive, and thus the result.   □

Remark. I had thought that this was a new result, but it turns out to be a result of Lebesgue. Furthermore, one can also show in this fashion that every positive integer is the sum of two squares and a triangular number.

8 Some More Applications

Let ω be a cube root of unity $\omega = \exp(2\pi i/3)$. We wish to begin by proving an elementary identity:

$$-\omega^2\theta^3\begin{bmatrix}\frac{1}{3}\\\frac{1}{3}\end{bmatrix}(\zeta, \tau) - \omega\theta^3\begin{bmatrix}\frac{1}{3}\\\frac{5}{3}\end{bmatrix}(\zeta, \tau) + \theta^3\begin{bmatrix}\frac{1}{3}\\1\end{bmatrix}(\zeta, \tau)$$

$$= c\theta\begin{bmatrix}\frac{1}{3}\\\frac{1}{3}\end{bmatrix}(\zeta, \tau)\theta\begin{bmatrix}\frac{1}{3}\\\frac{5}{3}\end{bmatrix}(\zeta, \tau)\theta\begin{bmatrix}\frac{1}{3}\\1\end{bmatrix}(\zeta, \tau),$$

where c is a constant independent of ζ. We say "elementary" identity because its proof simply requires showing that both sides of the equality have the same zeros. Then the quotient is an elliptic function with no poles and hence is constant. We will leave most of the arithmetic to the reader, but it is clear from the right-hand side that the zeros are the three points

$$\frac{\tau + 1}{3}, \quad \frac{\tau}{3}, \quad \frac{\tau + 2}{3}.$$

To see that $\zeta = \frac{\tau}{3}$ is a zero of the left-hand side we set $\zeta = \frac{\tau}{3}$, the zero of $\theta\begin{bmatrix}\frac{1}{3}\\1\end{bmatrix}(\zeta, \tau)$, in the equation, use equation (3), and obtain

$$-\omega^2\theta^3\begin{bmatrix}\frac{1}{3}\\\frac{1}{3}\end{bmatrix}\left(\frac{\tau}{3}, \tau\right) - \omega\theta\begin{bmatrix}\frac{1}{3}\\\frac{5}{3}\end{bmatrix}\left(\frac{\tau}{3}, \tau\right)$$

$$= -\omega^2\exp\left(6\pi i\begin{bmatrix}-1\ 2\ 1\\4\ 3\ 3\end{bmatrix}\right)\theta^3\begin{bmatrix}1\\\frac{1}{3}\end{bmatrix}(0, \tau) - \omega\exp\left(6\pi i\begin{bmatrix}-1\ 2\ 5\\4\ 3\ 3\end{bmatrix}\right)\theta^3\begin{bmatrix}1\\\frac{5}{3}\end{bmatrix}(0, \tau)$$

$$= \theta^3\begin{bmatrix}1\\\frac{1}{3}\end{bmatrix}(0, \tau) + \theta^3\begin{bmatrix}1\\\frac{5}{3}\end{bmatrix}(0, \tau) = 0.$$

In the same fashion we show that the other two points are also zeros of the left-hand side and thus can conclude that

$$\frac{-\omega^2\theta^3\begin{bmatrix}\frac{1}{3}\\\frac{1}{3}\end{bmatrix}(\zeta, \tau) - \omega\theta^3\begin{bmatrix}\frac{1}{3}\\\frac{5}{3}\end{bmatrix}(\zeta, \tau) + \theta^3\begin{bmatrix}\frac{1}{3}\\1\end{bmatrix}(\zeta, \tau)}{\theta\begin{bmatrix}\frac{1}{3}\\\frac{1}{3}\end{bmatrix}(\zeta, \tau)\theta\begin{bmatrix}\frac{1}{3}\\\frac{5}{3}\end{bmatrix}(\zeta, \tau)\theta\begin{bmatrix}\frac{1}{3}\\1\end{bmatrix}(\zeta, \tau)}$$

$$= \frac{-\omega^2\theta^3\begin{bmatrix}\frac{1}{3}\\\frac{1}{3}\end{bmatrix}(0, \tau) - \omega\theta^3\begin{bmatrix}\frac{1}{3}\\\frac{5}{3}\end{bmatrix}(0, \tau) + \theta^3\begin{bmatrix}\frac{1}{3}\\1\end{bmatrix}(0, \tau)}{\theta\begin{bmatrix}\frac{1}{3}\\\frac{1}{3}\end{bmatrix}(0, \tau)\theta\begin{bmatrix}\frac{1}{3}\\\frac{5}{3}\end{bmatrix}(0, \tau)\theta\begin{bmatrix}\frac{1}{3}\\1\end{bmatrix}(0, \tau)}.$$

If we now differentiate our identity and set $\zeta = \frac{\tau}{3}$, we obtain

$$\frac{-3\omega^2\theta^2 \begin{bmatrix} \frac{1}{3} \\ \frac{1}{3} \end{bmatrix}(\frac{\tau}{3},\tau)\,\theta'\begin{bmatrix} \frac{1}{3} \\ \frac{1}{3} \end{bmatrix}(\frac{\tau}{3},\tau) - 3\omega\theta^2\begin{bmatrix} \frac{1}{3} \\ \frac{5}{3} \end{bmatrix}(\frac{\tau}{3},\tau)}{\theta'\begin{bmatrix} \frac{1}{3} \\ 1 \end{bmatrix}(\frac{\tau}{3},\tau)\,\theta\begin{bmatrix} \frac{1}{3} \\ \frac{1}{3} \end{bmatrix}(\frac{\tau}{3},\tau)\,\theta\begin{bmatrix} \frac{1}{3} \\ \frac{5}{3} \end{bmatrix}(\frac{\tau}{3},\tau)}$$

$$= \frac{-\omega^2\theta^3\begin{bmatrix} \frac{1}{3} \\ \frac{1}{3} \end{bmatrix}(0,\tau) - \omega\theta^3\begin{bmatrix} \frac{1}{3} \\ \frac{5}{3} \end{bmatrix}(0,\tau) + \theta^3\begin{bmatrix} \frac{1}{3} \\ 1 \end{bmatrix}(0,\tau)}{\theta\begin{bmatrix} \frac{1}{3} \\ \frac{1}{3} \end{bmatrix}(0,\tau)\theta\begin{bmatrix} \frac{1}{3} \\ \frac{5}{3} \end{bmatrix}(0,\tau)\theta\begin{bmatrix} \frac{1}{3} \\ 1 \end{bmatrix}(0,\tau)}. \qquad (19)$$

In order to proceed, we need once again equation (3) and its companion, which is obtained by differentiation with respect to ζ. The equation is

$$\theta'\begin{bmatrix} \epsilon \\ \epsilon' \end{bmatrix}\left(\frac{n+m\tau}{2},\tau\right) = \exp\left(2\pi i\left[\frac{-m^2\tau}{8} - \frac{m(\epsilon'+n)}{4}\right]\right)$$

$$\times \left(\theta'\begin{bmatrix} \epsilon+m \\ \epsilon'+n \end{bmatrix}(0,\tau) - \pi im\theta\begin{bmatrix} \epsilon+m \\ \epsilon'+n \end{bmatrix}(0,\tau)\right). \quad (20)$$

Using equation (20), we find that the left-hand side of equation (19) becomes

$$-3\frac{\theta\begin{bmatrix} 1 \\ \frac{1}{3} \end{bmatrix}(0,\tau)\theta'\begin{bmatrix} 1 \\ \frac{1}{3} \end{bmatrix}(0,\tau)}{\theta\begin{bmatrix} 1 \\ \frac{5}{3} \end{bmatrix}(0,\tau)\theta'\begin{bmatrix} 1 \\ 1 \end{bmatrix}(0,\tau)} - 3\frac{\theta\begin{bmatrix} 1 \\ \frac{5}{3} \end{bmatrix}(0,\tau)\theta'\begin{bmatrix} 1 \\ \frac{5}{3} \end{bmatrix}(0,\tau)}{\theta\begin{bmatrix} 1 \\ \frac{1}{3} \end{bmatrix}(0,\tau)\theta'\begin{bmatrix} 1 \\ 1 \end{bmatrix}(0,\tau)}$$

$$+ 2\pi i\left(\frac{\theta^2\begin{bmatrix} 1 \\ \frac{1}{3} \end{bmatrix}(0,\tau)}{\theta\begin{bmatrix} 1 \\ \frac{5}{3} \end{bmatrix}(0,\tau)\theta'\begin{bmatrix} 1 \\ 1 \end{bmatrix}(0,\tau)} + \frac{\theta^2\begin{bmatrix} 1 \\ \frac{5}{3} \end{bmatrix}(0,\tau)}{\theta\begin{bmatrix} 1 \\ \frac{1}{3} \end{bmatrix}(0,\tau)\theta'\begin{bmatrix} 1 \\ 1 \end{bmatrix}(0,\tau)}\right).$$

This last expresion is clearly equal to

$$6\frac{\theta'\begin{bmatrix} 1 \\ \frac{1}{3} \end{bmatrix}(0,\tau)}{\theta'\begin{bmatrix} 1 \\ 1 \end{bmatrix}(0,\tau)},$$

since

$$\theta \begin{bmatrix} 1 \\ \frac{5}{3} \end{bmatrix} (0, \tau) = -\theta \begin{bmatrix} 1 \\ \frac{1}{3} \end{bmatrix} (0, \tau)$$

and

$$\theta' \begin{bmatrix} 1 \\ \frac{5}{3} \end{bmatrix} (0, \tau) = \theta' \begin{bmatrix} 1 \\ \frac{1}{3} \end{bmatrix} (0, \tau).$$

We state the result as a theorem.

Theorem 5. *For every τ in the upper half-plane we have*

$$6\theta' \begin{bmatrix} 1 \\ \frac{1}{3} \end{bmatrix} (0, \tau) \theta \begin{bmatrix} \frac{1}{3} \\ \frac{1}{3} \end{bmatrix} (0, \tau) \theta \begin{bmatrix} \frac{1}{3} \\ 1 \end{bmatrix} (0, \tau) \theta \begin{bmatrix} \frac{1}{3} \\ \frac{5}{3} \end{bmatrix} (0, \tau)$$

$$= \theta' \begin{bmatrix} 1 \\ 1 \end{bmatrix} (0, \tau)(-\omega^2 \theta^3 \begin{bmatrix} \frac{1}{3} \\ \frac{1}{3} \end{bmatrix} (0, \tau) - \omega \theta^3 \begin{bmatrix} \frac{1}{3} \\ \frac{5}{3} \end{bmatrix} (0, \tau) + \theta^3 \begin{bmatrix} \frac{1}{3} \\ 1 \end{bmatrix} (0, \tau)).$$

The above theorem has a nice interpertation in terms of infinite products. It
follows from the definitions that in the variable $x = \exp(2\pi i \tau / 3)$,

$$\theta' \begin{bmatrix} 1 \\ 1 \end{bmatrix} (0, \tau) = -2\pi x^{\frac{3}{8}} \sum_{n=0}^{\infty} (-1)^n (2n + 1) x^{\frac{3(n^2 + n)}{2}}$$

and

$$\theta \begin{bmatrix} \frac{1}{3} \\ 1 \end{bmatrix} (0, \tau) = \exp(\pi i / 6) x^{\frac{1}{24}} \sum_{n=-\infty}^{n=\infty} (-1)^n x^{\frac{3n^2 + n}{2}}.$$

By Jacobi's identity,

$$\theta^3 \begin{bmatrix} \frac{1}{3} \\ 1 \end{bmatrix} (0, \tau) = i x^{\frac{1}{8}} \sum_{n=0}^{\infty} (-1)^n (2n + 1) x^{\frac{n^2 + n}{2}}$$

and thus

$$\theta^3 \begin{bmatrix} \frac{1}{3} \\ 1 \end{bmatrix} (0, 3\tau) = i x^{\frac{3}{8}} \sum_{n=0}^{\infty} (-1)^n (2n + 1) x^{\frac{3(n^2 + n)}{2}}.$$

The Jacobi triple product formula and a little thought then yields (see [5]) that

$$\theta \begin{bmatrix} \frac{1}{3} \\ \frac{1}{3} \end{bmatrix} (0, \tau) = \exp \left(\frac{\pi i}{18} \right) x^{\frac{1}{24}} \prod_{n=1}^{\infty} (1 - (\omega x)^n),$$

$$\theta \begin{bmatrix} \frac{1}{3} \\ \frac{5}{3} \end{bmatrix} (0, \tau) = \exp \left(\frac{5\pi i}{18} \right) x^{\frac{1}{24}} \prod_{n=1}^{\infty} (1 - (\omega^2 x)^n),$$

and

$$\theta\begin{bmatrix}\frac{1}{3}\\1\end{bmatrix}(0,\tau) = \exp\left(\frac{\pi i}{6}\right)x^{\frac{1}{24}}\prod_{n=1}^{\infty}(1-x^n).$$

It thus follows that

$$\theta^3\begin{bmatrix}\frac{1}{3}\\1\end{bmatrix}(0,3\tau) = ix^{\frac{3}{8}}\prod_{n=1}^{\infty}(1-x^{3n})^3$$

and that

$$2\pi i\theta^3\begin{bmatrix}\frac{1}{3}\\1\end{bmatrix}(0,3\tau) = -2\pi x^{\frac{3}{8}}\prod_{n=1}^{\infty}(1-x^{3n})^3.$$

Finally, we observe that

$$\theta\begin{bmatrix}\frac{1}{3}\\\frac{1}{3}\end{bmatrix}(0,\tau)\theta\begin{bmatrix}\frac{1}{3}\\1\end{bmatrix}(0,\tau)\theta\begin{bmatrix}\frac{1}{3}\\\frac{5}{3}\end{bmatrix}(0,\tau)$$

$$= ix^{\frac{1}{8}}\prod_{n=1}^{\infty}(1-x^n)(1-(\omega x)^n)(1-(\omega^2 x)^n).$$

For each n, $(1-x^n)(1-(\omega x)^n)(1-(\omega^2 x)^n)$ is equal to $(1-x^n)^3$ for n congruent to zero mod three and is equal to $(1-x^{3n})$ otherwise. It thus follows that

$$\prod_{n=1}^{\infty}(1-x^n)(1-(\omega x)^n)(1-(\omega^2 x)^n)$$

$$= \prod_{k=0}^{\infty}(1-x^{9k+3})(1-x^{9k+6})(1-x^{3k+3})^3.$$

It thus follows that

$$\frac{\theta'\begin{bmatrix}1\\1\end{bmatrix}(0,\tau)}{\theta\begin{bmatrix}\frac{1}{3}\\\frac{1}{3}\end{bmatrix}(0,\tau)\theta\begin{bmatrix}\frac{1}{3}\\1\end{bmatrix}(0,\tau)\theta\begin{bmatrix}\frac{1}{3}\\\frac{5}{3}\end{bmatrix}(0,\tau)} = \frac{2\pi i x^{\frac{1}{4}}}{\prod_{k=0}^{\infty}(1-x^{9k+3})(1-x^{9k+6})}.$$

Corollary 4. *In the variable* $x = \exp(2\pi i\tau)$ *we have*

$$\frac{6\theta'\begin{bmatrix}1\\\frac{1}{3}\end{bmatrix}(0,\tau)}{-\omega^2\theta^3\begin{bmatrix}\frac{1}{3}\\\frac{1}{3}\end{bmatrix}(0,\tau) - \omega\theta^3\begin{bmatrix}\frac{1}{3}\\\frac{5}{3}\end{bmatrix}(0,\tau) + \theta^3\begin{bmatrix}\frac{1}{3}\\1\end{bmatrix}(0,\tau)}$$

$$= \frac{2\pi i x^{\frac{1}{4}}}{\prod_{n=0}^{\infty}(1-x^{3k+1})(1-x^{3k+2})}.$$

If we divide the right-hand side of the equation in the corollary above by $2\pi i x^{\frac{1}{4}}$, the right-hand side has an interesting interpretation. The coefficient of x^n in the power series expansion is the number of ways you can write n using summands that are congruent to 1 and 2 mod three. We sometimes call this the number of partitions of n from the set of positive integers congruent to 1 and 2 mod three. This has led us to another subject, which we shall treat in the next section. Before we do this, however, we derive a further theta function identity.

We wish to find complex numbers x_i, $i = 1, \ldots, 4$, not all zero such that

$$x_1\theta^3\begin{bmatrix}\frac{1}{3}\\\frac{1}{3}\end{bmatrix}(\zeta,\tau) + x_2\theta^3\begin{bmatrix}\frac{1}{3}\\1\end{bmatrix}(\zeta,\tau) + x_3\theta^3\begin{bmatrix}\frac{1}{3}\\\frac{5}{3}\end{bmatrix}(\zeta,\tau) + x_4\theta^3\begin{bmatrix}1\\\frac{1}{3}\end{bmatrix}(\zeta,\tau) = 0.$$

The reason we can do this is that the four functions are linearly dependent over the complex numbers. This fact is a simple exercise, which we leave for the reader.

In order to find the numbers x_i we substitute successively $\zeta = \frac{\tau+1}{3}, \frac{\tau}{3}, \frac{\tau-1}{3}, \frac{1}{3}$ and obtain a system of four homogeneous linear equations for the unknowns x_i. This system has a nontrivial solution, and thus the determinant of the system vanishes. Moreover, one can check that the rank of the matrix is 3, so that there is a one-dimensional solution space. A basis for this space is the minors of the last row of the matrix, and one can compute the basis vector as

$$x_1 = \theta^3\begin{bmatrix}\frac{1}{3}\\\frac{1}{3}\end{bmatrix}(0,\tau)\theta^3\begin{bmatrix}\frac{1}{3}\\\frac{5}{3}\end{bmatrix}(0,\tau)(\omega - \omega^2),$$

$$x_2 = \theta^3\begin{bmatrix}\frac{1}{3}\\\frac{1}{3}\end{bmatrix}(0,\tau)\theta^3\begin{bmatrix}\frac{1}{3}\\1\end{bmatrix}(0,\tau)(1 - \omega),$$

$$x_3 = \theta^3\begin{bmatrix}\frac{1}{3}\\\frac{5}{3}\end{bmatrix}(0,\tau)\theta^3\begin{bmatrix}\frac{1}{3}\\1\end{bmatrix}(0,\tau)(1 - \omega^2),$$

$$x_4 = \theta^3\begin{bmatrix}1\\\frac{1}{3}\end{bmatrix}(0,\tau)(\theta^3\begin{bmatrix}\frac{1}{3}\\1\end{bmatrix}(0,\tau) - \omega\theta^3\begin{bmatrix}\frac{1}{3}\\\frac{5}{3}\end{bmatrix}(0,\tau) - \omega^2\theta^3\begin{bmatrix}\frac{1}{3}\\\frac{1}{3}\end{bmatrix}(0,\tau)).$$

We have therefore proved the following identity.

Theorem 6. *For all τ in the upper half-plane we have*

$$(\omega - \omega^2)\theta^3\begin{bmatrix}\frac{1}{3}\\\frac{1}{3}\end{bmatrix}(0,\tau)\theta^3\begin{bmatrix}\frac{1}{3}\\\frac{5}{3}\end{bmatrix}(0,\tau)\theta^3\begin{bmatrix}\frac{1}{3}\\\frac{1}{3}\end{bmatrix}(\zeta,\tau)$$

$$+ (1 - \omega)\theta^3\begin{bmatrix}\frac{1}{3}\\\frac{1}{3}\end{bmatrix}(0,\tau)\theta^3\begin{bmatrix}\frac{1}{3}\\1\end{bmatrix}(0,\tau)\theta^3\begin{bmatrix}\frac{1}{3}\\1\end{bmatrix}(\zeta,\tau)$$

$$+ (1 - \omega^2)\theta^3\begin{bmatrix}\frac{1}{3}\\\frac{5}{3}\end{bmatrix}(0,\tau)\theta^3\begin{bmatrix}\frac{1}{3}\\1\end{bmatrix}(0,\tau)\theta^3\begin{bmatrix}\frac{1}{3}\\\frac{5}{3}\end{bmatrix}(\zeta,\tau)$$

$$= \theta^3 \begin{bmatrix} 1 \\ \frac{1}{3} \end{bmatrix}(0, \tau)(\omega^2\theta^3 \begin{bmatrix} \frac{1}{3} \\ \frac{1}{3} \end{bmatrix}(0, \tau) + \omega\theta^3 \begin{bmatrix} \frac{1}{3} \\ \frac{5}{3} \end{bmatrix}(0, \tau)$$

$$- \theta^3 \begin{bmatrix} \frac{1}{3} \\ 1 \end{bmatrix}(0, \tau))\theta^3 \begin{bmatrix} 1 \\ \frac{1}{3} \end{bmatrix}(\zeta, \tau).$$

In particular, if we replace ζ by $\zeta + \frac{1}{3}$ and use equation (3), we get

$$(\omega - \omega^2)\theta^3 \begin{bmatrix} \frac{1}{3} \\ \frac{1}{3} \end{bmatrix}(0, \tau)\theta^3 \begin{bmatrix} \frac{1}{3} \\ \frac{5}{3} \end{bmatrix}(0, \tau)\theta^3 \begin{bmatrix} \frac{1}{3} \\ 1 \end{bmatrix}(\zeta, \tau)$$

$$+ (1 - \omega)\theta^3 \begin{bmatrix} \frac{1}{3} \\ \frac{1}{3} \end{bmatrix}(0, \tau)\theta^3 \begin{bmatrix} \frac{1}{3} \\ 1 \end{bmatrix}(0, \tau)\theta^3 \begin{bmatrix} \frac{1}{3} \\ \frac{5}{3} \end{bmatrix}(\zeta, \tau)$$

$$+ (1 - \omega^2)\theta^3 \begin{bmatrix} \frac{1}{3} \\ \frac{5}{3} \end{bmatrix}(0, \tau)\theta^3 \begin{bmatrix} \frac{1}{3} \\ 1 \end{bmatrix}(0, \tau)\theta^3 \begin{bmatrix} \frac{1}{3} \\ \frac{7}{3} \end{bmatrix}(\zeta, \tau)$$

$$= \theta^3 \begin{bmatrix} 1 \\ \frac{1}{3} \end{bmatrix}(0, \tau)(\omega^2\theta^3 \begin{bmatrix} \frac{1}{3} \\ \frac{1}{3} \end{bmatrix}(0, \tau) + \omega\theta^3 \begin{bmatrix} \frac{1}{3} \\ \frac{5}{3} \end{bmatrix}(0, \tau)$$

$$- \theta^3 \begin{bmatrix} \frac{1}{3} \\ 1 \end{bmatrix}(0, \tau))\theta^3 \begin{bmatrix} 1 \\ 1 \end{bmatrix}(\zeta, \tau).$$

It follows from the above identity, which of course by other changes of the variable ζ gives many more, that the left-hand side of the identity vanishes to order 3 in ζ. Hence the left-hand side and its first two derivatives vanish at the origin. The vanishing of the left-hand side at the origin gives us the equation

$$(\omega - \omega^2)\theta^3 \begin{bmatrix} \frac{1}{3} \\ \frac{1}{3} \end{bmatrix}(0, \tau)\theta^3 \begin{bmatrix} \frac{1}{3} \\ \frac{5}{3} \end{bmatrix}(0, \tau)\theta^3 \begin{bmatrix} \frac{1}{3} \\ 1 \end{bmatrix}(0, \tau)$$

$$+ (1 - \omega)\theta^3 \begin{bmatrix} \frac{1}{3} \\ \frac{1}{3} \end{bmatrix}(0, \tau)\theta^3 \begin{bmatrix} \frac{1}{3} \\ 1 \end{bmatrix}(0, \tau)\theta^3 \begin{bmatrix} \frac{1}{3} \\ \frac{5}{3} \end{bmatrix}(0, \tau)$$

$$- (1 - \omega^2)\theta^3 \begin{bmatrix} \frac{1}{3} \\ 1 \end{bmatrix}(0, \tau)\theta^3 \begin{bmatrix} \frac{1}{3} \\ \frac{5}{3} \end{bmatrix}(0, \tau)\theta^3 \begin{bmatrix} \frac{1}{3} \\ \frac{1}{3} \end{bmatrix}(0, \tau) = 0.$$

The last equation follows from the fact that $\theta^3 \begin{bmatrix} \frac{1}{3} \\ \frac{1}{3} \end{bmatrix}(0, \tau) = -\theta^3 \begin{bmatrix} \frac{1}{3} \\ \frac{7}{3} \end{bmatrix}(0, \tau)$.

This is of course a trivial identity, since clearly

$$(\omega - \omega^2) + (1 - \omega) - (1 - \omega^2) = 0.$$

We turn now to the first derivative of the left-hand side and evaluate at the origin. We obtain

$$
\theta^3 \begin{bmatrix} \frac{1}{3} \\ \frac{1}{3} \end{bmatrix} (0, \tau) \theta^3 \begin{bmatrix} \frac{1}{3} \\ \frac{5}{3} \end{bmatrix} (0, \tau) \theta^3 \begin{bmatrix} \frac{1}{3} \\ 1 \end{bmatrix} (0, \tau)
$$

$$
\times \left[(\omega - \omega^2) \frac{\theta' \begin{bmatrix} \frac{1}{3} \\ 1 \end{bmatrix} (0, \tau)}{\theta \begin{bmatrix} \frac{1}{3} \\ 1 \end{bmatrix} (0, \tau)} + (1 - \omega) \frac{\theta' \begin{bmatrix} \frac{1}{3} \\ \frac{5}{3} \end{bmatrix} (0, \tau)}{\theta \begin{bmatrix} \frac{1}{3} \\ \frac{5}{3} \end{bmatrix} (0, \tau)} - (1 - \omega^2) \frac{\theta' \begin{bmatrix} \frac{1}{3} \\ \frac{1}{3} \end{bmatrix} (0, \tau)}{\theta \begin{bmatrix} \frac{1}{3} \\ 1 \end{bmatrix} (0, \tau)} \right].
$$

The vanishing of this derivative gives us, after dividing by $(1 - \omega)$ and multiplying by ω^2, the identity

$$
\frac{\theta' \begin{bmatrix} \frac{1}{3} \\ 1 \end{bmatrix} (0, \tau)}{\theta \begin{bmatrix} \frac{1}{3} \\ 1 \end{bmatrix} (0, \tau)} + \omega^2 \frac{\theta' \begin{bmatrix} \frac{1}{3} \\ \frac{5}{3} \end{bmatrix} (0, \tau)}{\theta \begin{bmatrix} \frac{1}{3} \\ \frac{5}{3} \end{bmatrix} (0, \tau)} + \omega \frac{\theta' \begin{bmatrix} \frac{1}{3} \\ \frac{1}{3} \end{bmatrix} (0, \tau)}{\theta \begin{bmatrix} \frac{1}{3} \\ \frac{1}{3} \end{bmatrix} (0, \tau)} = 0. \tag{21}
$$

The above identity is quite elegant, and one wonders what it tells us.

Theorem 7. *The combinatoric content of the above identity is the statement that if N is congruent to 2 mod 3, then the number of divisors of N congruent to 1 mod 3 equals the number of divisors of N congruent to 2 mod 3.*

Proof. The proof uses the Jacobi triple product formula and logarithmic differentiation. We have by Jacobi in the variable $y = e^{\frac{2\pi i \tau}{3}}$ that

$$
\frac{\theta' \begin{bmatrix} \frac{1}{3} \\ \frac{1}{3} \end{bmatrix} (0, \tau)}{\theta \begin{bmatrix} \frac{1}{3} \\ \frac{1}{3} \end{bmatrix} (0, \tau)} = (2\pi i) \left[\frac{1}{6} + \sum_{n=0}^{\infty} \frac{e^{\frac{-4\pi i}{3}} y^{3n+1}}{1 - e^{\frac{-4\pi i}{3}} y^{3n+1}} - \sum_{n=0}^{\infty} \frac{e^{\frac{4\pi i}{3}} y^{3n+2}}{1 - e^{\frac{4\pi i}{3}} y^{3n+2}} \right],
$$

$$
\frac{\theta' \begin{bmatrix} \frac{1}{3} \\ \frac{5}{3} \end{bmatrix} (0, \tau)}{\theta \begin{bmatrix} \frac{1}{3} \\ \frac{5}{3} \end{bmatrix} (0, \tau)} = (2\pi i) \left[\frac{1}{6} + \sum_{n=0}^{\infty} \frac{e^{\frac{-2\pi i}{3}} y^{3n+1}}{1 - e^{\frac{-2\pi i}{3}} y^{3n+1}} - \sum_{n=0}^{\infty} \frac{e^{\frac{2\pi i}{3}} y^{3n+2}}{1 - e^{\frac{2\pi i}{3}} y^{3n+2}} \right],
$$

$$\frac{\theta'\begin{bmatrix}\frac{1}{3}\\1\end{bmatrix}(0,\tau)}{\theta\begin{bmatrix}\frac{1}{3}\\1\end{bmatrix}(0,\tau)} = (2\pi i)\left[\frac{1}{6} + \sum_{n=0}^{\infty}\frac{y^{3n+1}}{1-y^{3n+1}} - \sum_{n=0}^{\infty}\frac{y^{3n+2}}{1-y^{3n+2}}\right].$$

We now write each of the above sums as a power series in y and obtain successively

$$(2\pi i)\left[\frac{1}{6} + \sum_{n=0}^{\infty}A_n y^n\right],$$

where $A_n = \sum_{d|n,d\equiv 1}\omega^{\frac{n}{d}} - \sum_{d|n,d\equiv 2}\omega^{\frac{2n}{d}}$;

$$(2\pi i)\left[\frac{1}{6} + \sum_{n=0}^{\infty}B_n y^n\right],$$

where $B_n = \sum_{d|n,d\equiv 1}\omega^{\frac{2n}{d}} - \sum_{d|n,d\equiv 2}\omega^{\frac{n}{d}}$; and

$$(2\pi i)\left[\frac{1}{6} + \sum_{n=0}^{\infty}C_n y^n\right],$$

where $C_n = \sum_{d|n,d\equiv 1}1 - \sum_{d|n,d\equiv 2}1$. It is quite clear now that C_n is simply the difference between the number of divisors of N congruent to 1 mod 3 and those congruent to 2 mod 3. The numbers A_n, B_n, however, depend on the congruence class of N mod 3.

It is easy to see that

$$A_n = \begin{Bmatrix} n=0 & r-s \\ n=1 & \omega(r-s) \\ n=2 & \omega^2(r-s) \end{Bmatrix}, \quad B_n = \begin{Bmatrix} n=0 & r-s \\ n=1 & \omega^2(r-s) \\ n=2 & \omega(r-s) \end{Bmatrix},$$

where r is the number of divisors of N congruent to 1 mod 3 and s is the number of divisors of N congruent to 2 mod 3.

The identity obtained above from the vanishing of the first derivative at the origin in terms of the complex numbers A_n, B_n, C_n is thus given by $C_n + \omega^2 B_n + \omega A_n = 0$, and this gives

$$n \equiv 0 \quad (r-s)(1+\omega+\omega^2),$$
$$n \equiv 1 \quad (r-s)(1+\omega+\omega^2),$$
$$n \equiv 2 \quad 3(r-s).$$

It thus follows that the vanishing of the first derivative is equivalent to the statement that when $n \equiv 2$ mod 3, the number of divisors of n congruent to 1 mod 3 equals the number of divisors of n congruent to 2 mod 3.   □

For more on this subject see [1] and [8].

We now turn our attention to the second derivative at the origin. The identity we get here once again after dividing by $(1 - \omega)$ and multiplying by ω^2 is the following:

$$
\left(\frac{\theta'' \begin{bmatrix} \frac{1}{3} \\ 1 \end{bmatrix} (0, \tau)}{\theta \begin{bmatrix} \frac{1}{3} \\ 1 \end{bmatrix} (0, \tau)} + 2 \left(\frac{\theta' \begin{bmatrix} \frac{1}{3} \\ 1 \end{bmatrix} (0, \tau)}{\theta \begin{bmatrix} \frac{1}{3} \\ 1 \end{bmatrix} (0, \tau)} \right)^2 \right)
$$

$$
+ \omega^2 \left(\frac{\theta'' \begin{bmatrix} \frac{1}{3} \\ \frac{5}{3} \end{bmatrix} (0, \tau)}{\theta \begin{bmatrix} \frac{1}{3} \\ \frac{5}{3} \end{bmatrix} (0, \tau)} + 2 \left(\frac{\theta' \begin{bmatrix} \frac{1}{3} \\ \frac{5}{3} \end{bmatrix} (0, \tau)}{\theta \begin{bmatrix} \frac{1}{3} \\ \frac{5}{3} \end{bmatrix} (0, \tau)} \right)^2 \right)
$$

$$
+ \omega \left(\frac{\theta'' \begin{bmatrix} \frac{1}{3} \\ \frac{1}{3} \end{bmatrix} (0, \tau)}{\theta \begin{bmatrix} \frac{1}{3} \\ \frac{1}{3} \end{bmatrix} (0, \tau)} + 2 \left(\frac{\theta' \begin{bmatrix} \frac{1}{3} \\ \frac{1}{3} \end{bmatrix} (0, \tau)}{\theta \begin{bmatrix} \frac{1}{3} \\ \frac{1}{3} \end{bmatrix} (0, \tau)} \right)^2 \right) = 0. \tag{22}
$$

We would now like to understand what this identity says combinatorially. Our first simplification is by use of the heat equation, equation (5), to rewrite equation (23) as

$$
4\pi i \left[\frac{d/d\tau \theta \begin{bmatrix} \frac{1}{3} \\ 1 \end{bmatrix} (0, \tau)}{\theta \begin{bmatrix} \frac{1}{3} \\ 1 \end{bmatrix} (0, \tau)} + \omega^2 \frac{d/d\tau \theta \begin{bmatrix} \frac{1}{3} \\ \frac{5}{3} \end{bmatrix} (0, \tau)}{\theta \begin{bmatrix} \frac{1}{3} \\ \frac{5}{3} \end{bmatrix} (0, \tau)} + \omega \frac{d/d\tau \theta \begin{bmatrix} \frac{1}{3} \\ \frac{1}{3} \end{bmatrix} (0, \tau)}{\theta \begin{bmatrix} \frac{1}{3} \\ \frac{1}{3} \end{bmatrix} (0, \tau)} \right]
$$

$$
+ 2 \left[\left(\frac{\theta' \begin{bmatrix} \frac{1}{3} \\ 1 \end{bmatrix} (0, \tau)}{\theta \begin{bmatrix} \frac{1}{3} \\ 1 \end{bmatrix} (0, \tau)} \right)^2 + \omega^2 \left(\frac{\theta' \begin{bmatrix} \frac{1}{3} \\ \frac{5}{3} \end{bmatrix} (0, \tau)}{\theta \begin{bmatrix} \frac{1}{3} \\ \frac{5}{3} \end{bmatrix} (0, \tau)} \right)^2 + \omega \left(\frac{\theta' \begin{bmatrix} \frac{1}{3} \\ \frac{1}{3} \end{bmatrix} (0, \tau)}{\theta \begin{bmatrix} \frac{1}{3} \\ \frac{1}{3} \end{bmatrix} (0, \tau)} \right)^2 \right]
$$

$$
= 0.
$$

By virtue of equation (21) we can rewrite the above as

$$
4\pi i \left[\frac{d/d\tau\theta \begin{bmatrix} \frac{1}{3} \\ 1 \end{bmatrix}(0,\tau)}{\theta \begin{bmatrix} \frac{1}{3} \\ 1 \end{bmatrix}(0,\tau)} + \omega^2 \frac{d/d\tau\theta \begin{bmatrix} \frac{1}{3} \\ \frac{5}{3} \end{bmatrix}(0,\tau)}{\theta \begin{bmatrix} \frac{1}{3} \\ \frac{5}{3} \end{bmatrix}(0,\tau)} + \omega \frac{d/d\tau\theta \begin{bmatrix} \frac{1}{3} \\ \frac{1}{3} \end{bmatrix}(0,\tau)}{\theta \begin{bmatrix} \frac{1}{3} \\ \frac{1}{3} \end{bmatrix}(0,\tau)} \right]
$$

$$
-2 \left(\frac{\theta' \begin{bmatrix} \frac{1}{3} \\ \frac{5}{3} \end{bmatrix}(0,\tau)}{\theta \begin{bmatrix} \frac{1}{3} \\ \frac{5}{3} \end{bmatrix}(0,\tau)} - \frac{\theta' \begin{bmatrix} \frac{1}{3} \\ \frac{1}{3} \end{bmatrix}(0,\tau)}{\theta \begin{bmatrix} \frac{1}{3} \\ \frac{1}{3} \end{bmatrix}(0,\tau)} \right)^2 = 0.
$$

Theorem 8. *Let $\delta(n)$ be the difference between the number of divisors of n congruent to 1 mod 3 and the number of divisors of n congruent to 2 mod 3. Let $\sigma(n)$ denote the sum of the divisors of n. The combinatorial content of the above identity is that*

$$
\sum_{n=0}^{\infty} \sigma(3n+2)y^{3n+2} = 3 \left(\sum_{n=0}^{\infty} \delta(3n+1)y^{3n+1} \right)^2,
$$

or that

$$
\sigma(3n+2) = 3 \sum_{k=0}^{n} \delta(3k+1)\delta(3(n-k)+1).
$$

Proof. The proof is much like the proof of Theorem 7. The Jacobi triple product formula gives

$$
\frac{d/d\tau\theta \begin{bmatrix} \frac{1}{3} \\ \frac{1}{3} \end{bmatrix}(0,\tau)}{\theta \begin{bmatrix} \frac{1}{3} \\ \frac{1}{3} \end{bmatrix}(0,\tau)} = \frac{2\pi i}{3} \left[\frac{1}{24} + \sum_{n=0}^{\infty} \frac{-(3n+1)e^{\frac{2\pi i}{3}}y^{3n+1}}{1-e^{\frac{2\pi i}{3}}y^{3n+1}} \right.
$$

$$
\left. + \sum_{n=0}^{\infty} \frac{-(3n+2)e^{\frac{4\pi i}{3}}y^{3n+2}}{1-e^{\frac{4\pi i}{3}}y^{3n+2}} + \sum_{n=0}^{\infty} \frac{-(3n+3)y^{3n+3}}{1-y^{3n+3}} \right],
$$

$$
\frac{d/d\tau\theta \begin{bmatrix} \frac{1}{3} \\ \frac{5}{3} \end{bmatrix}(0,\tau)}{\theta \begin{bmatrix} \frac{1}{3} \\ \frac{5}{3} \end{bmatrix}(0,\tau)} = \frac{2\pi i}{3} \left[\frac{1}{24} + \sum_{n=0}^{\infty} \frac{-(3n+1)e^{\frac{4\pi i}{3}}y^{3n+1}}{1-e^{\frac{4\pi i}{3}}y^{3n+1}} \right.
$$

$$
\left. + \sum_{n=0}^{\infty} \frac{-(3n+2)e^{\frac{2\pi i}{3}}y^{3n+2}}{1-e^{\frac{2\pi i}{3}}y^{3n+2}} + \sum_{n=0}^{\infty} \frac{-(3n+3)y^{3n+3}}{1-y^{3n+3}} \right],
$$

and

$$\frac{d/d\tau\theta\begin{bmatrix}\frac{1}{3}\\1\end{bmatrix}(0,\tau)}{\theta\begin{bmatrix}\frac{1}{3}\\1\end{bmatrix}(0,\tau)} = \frac{2\pi i}{3}\left[\frac{1}{24} + \sum_{n=0}^{\infty}\frac{-(3n+1)y^{3n+1}}{1-y^{3n+1}}\right.$$

$$\left. + \sum_{n=0}^{\infty}\frac{-(3n+2)y^{3n+2}}{1-y^{3n+2}} + \sum_{n=0}^{\infty}\frac{-(3n+3)y^{3n+3}}{1-y^{3n+3}}\right].$$

Once again as in the proof of Theorem 7 we write each of the above as a power series and obtain

$$\frac{d/d\tau\theta\begin{bmatrix}\frac{1}{3}\\\frac{1}{3}\end{bmatrix}(0,\tau)}{\theta\begin{bmatrix}\frac{1}{3}\\\frac{1}{3}\end{bmatrix}(0,\tau)} = \frac{2\pi i}{3}\left[\frac{1}{24} + \sum_{n=0}^{\infty}\tilde{A}_n y^n\right],$$

where

$$\tilde{A}_n = \begin{cases} n \equiv 0 & -\sigma(n) \\ n \equiv 1 & -\omega\left(\sum_{d|n,d\equiv 1} d + \sum_{d|n,d\equiv 2} d\right) \\ n \equiv 2 & -\omega^2\left(\sum_{d|n,d\equiv 1} d + \sum_{d|n,d\equiv 2} d\right) \end{cases}$$

$$\frac{d/d\tau\theta\begin{bmatrix}\frac{1}{3}\\\frac{5}{3}\end{bmatrix}(0,\tau)}{\theta\begin{bmatrix}\frac{1}{3}\\\frac{5}{3}\end{bmatrix}(0,\tau)} = \frac{2\pi i}{3}\left[\frac{1}{24} + \sum_{n=0}^{\infty}\tilde{B}_n y^n\right],$$

where

$$\tilde{B}_n = \begin{cases} n \equiv 0 & -\sigma(n) \\ n \equiv 1 & -\omega^2\left(\sum_{d|n,d\equiv 1} d + \sum_{d|n,d\equiv 2} d\right) \\ n \equiv 2 & -\omega\left(\sum_{d|n,d\equiv 1} d + \sum_{d|n,d\equiv 2} d\right) \end{cases}.$$

and finally,

$$\frac{d/d\tau\theta\begin{bmatrix}\frac{1}{3}\\1\end{bmatrix}(0,\tau)}{\theta\begin{bmatrix}\frac{1}{3}\\1\end{bmatrix}(0,\tau)} = \frac{2\pi i}{3}\left[\frac{1}{24} + \sum_{n=0}^{\infty}\tilde{C}_n y^n\right],$$

where

$$\tilde{C}_n = -\sigma(n).$$

It thus follows that

$$
\frac{d/d\tau\theta\begin{bmatrix}\frac{1}{3}\\1\end{bmatrix}(0,\tau)}{\theta\begin{bmatrix}\frac{1}{3}\\1\end{bmatrix}(0,\tau)}+\omega\frac{d/d\tau\theta\begin{bmatrix}\frac{1}{3}\\\frac{1}{3}\end{bmatrix}(0,\tau)}{\theta\begin{bmatrix}\frac{1}{3}\\\frac{1}{3}\end{bmatrix}(0,\tau)}+\omega^2\frac{d/d\tau\theta\begin{bmatrix}\frac{1}{3}\\\frac{5}{3}\end{bmatrix}(0,\tau)}{\theta\begin{bmatrix}\frac{1}{3}\\\frac{5}{3}\end{bmatrix}(0,\tau)}
$$

$$
=\frac{2\pi i}{3}\left[\sum_{n=1}^{\infty}(\tilde{C}_n+\omega\tilde{A}_n+\omega^2\tilde{B}_n)y^n\right]=-2\pi i\sum_{n=0}^{\infty}\sigma(3n+2)y^{3n+2}.
$$

The identity of equation (22) and its rewrites now become

$$
8\pi^2\sum_{n=0}^{\infty}\sigma(3n+2)y^{3n+2}=2\left(\frac{\theta'\begin{bmatrix}\frac{1}{3}\\\frac{5}{3}\end{bmatrix}(0,\tau)}{\theta\begin{bmatrix}\frac{1}{3}\\\frac{5}{3}\end{bmatrix}(0,\tau)}-\frac{\theta'\begin{bmatrix}\frac{1}{3}\\\frac{1}{3}\end{bmatrix}(0,\tau)}{\theta\begin{bmatrix}\frac{1}{3}\\\frac{1}{3}\end{bmatrix}(0,\tau)}\right)^2
$$

$$
=24\pi^2\left(\sum_{n=0}^{\infty}\delta(3n+1)y^{3n+1}\right)^2.
$$

Finally, the identity just says that

$$
\sum_{n=0}^{\infty}\sigma(3n+2)y^n=3\left(\sum_{n=0}^{\infty}\delta(3n+1)y^n\right)^2.
$$

□

For another proof of this identity see [1] and [2].

We remark that the purpose of this lengthy computation has been to show how simply studying the theta function leads one to significant combinatorial results. We also should remark that the result of Theorem 7 is really rather elementary and surely does not need theta function theory. The statement that $r = s$ or that when a number is congruent to 2 mod 3 the number of divisors congruent to 1 mod 3 equals the number of divisors congruent to 2 mod 3 is extremely elementary. It follows from the obvious fact that there is a bijective map of the divisors congruent to 1 mod 3 onto those congruent to 2 mod 3. The map is just $d \mapsto \frac{n}{d}$. When n is congruent to 2 mod 3 this is clearly a bijective map.

9 Partitions of the Positive Integers

Let S be a set of positive integers and let N be an integer. A partition of N from S is a sum of integers in S that add up to N, where we order the summands according to size. For example, if S is the entire set of integers and $N = 4$, then the partitions of 4 are

$$4, \quad 3 + 1, \quad 2 + 2, \quad 2 + 1 + 1, \quad 1 + 1 + 1 + 1.$$

It is easy to see that if we denote by $P(n)$ the number of partitions of the positive integer n in the set of positive integers, then

$$\frac{1}{\prod_{n=1}^{\infty}(1 - x^n)} = 1 + \sum_{n=1}^{\infty} P(n)x^n.$$

When dealing with partitions from a given set S, we can place restrictions on the type of partitions we shall permit. For example, we can ask that the partition not have any repetitions. In the example above, if we demand no repetitions, then only the first two partitions are valid, since the remaining three partitions all contain repetitions. Perhaps the best-known example is the following:

Let S be the set of positive integers and T the set of positive odd integers. Then the number of partitions of n from S without repetitions equals the number of partitions of n from T with no restriction. If we denote the former by $P(n, \cdot S)$ and the latter by $P(n, T)$, then the statement is that $P(n, s) = P(n, T)$. The proof is the following obvious identity:

$$\sum_{n=0}^{\infty} P(n, S)x^n = \prod_{n=1}^{\infty}(1 + x^n) = \frac{1}{\prod_{n=0}^{\infty}(1 - x^{2n-1})} = \sum_{n=0}^{\infty} P(n, T)x^n.$$

In [10], Andrews asked the following question:

For what sets of positive integers S, T is it true that $P(n, S) = P(n - 1, T)$? Andrews points out that there is a master identity that partially answers this question. In what follows we shall show that in fact, a slight variant of equation (4) will give the solutions discussed by Andrews. We learned of this question from Frank Garvan, who also provided us with [11].

It follows from equation (4) that for $0 < l < N$ and $N > 1$,

$$\theta \begin{bmatrix} \frac{l}{N} \\ 1 \end{bmatrix} (0, N\tau) = \theta \begin{bmatrix} \frac{l}{2N} \\ 2 \end{bmatrix} (0, 4N\tau) + \theta \begin{bmatrix} \frac{l-2N}{2N} \\ 2 \end{bmatrix} (0, 4N\tau),$$

and therefore by the Jacobi triple product formula of Section 5, we have, after cancellation,

$$\prod_{n=0}^{\infty}(1 - x^{2N(n+1)})(1 - x^{2Nn+N+l})(1 - x^{2Nn+N-l})$$

$$= \prod_{n=0}^{\infty}(1 - x^{8N(n+1)})(1 + x^{8Nn+4N+2l})(1 + x^{8Nn+4N-2l})$$

$$- x \prod_{n=0}^{\infty}(1 - x^{8N(n+1)})(1 + x^{8Nn+2l})(1 + x^{8Nn+8N-2l}).$$

If we now take the examples $N = 2, l = 1, N = 5, l = 1, l = 3$, we shall obtain the three shifted identities obtained with one master identity by Andrews in [10]. In a sense, our result is really not different from Andrews' but seems much more natural than the mysterious yet simple identity used by Andrews.

The case $N = 2, l = 1$ yields the identity

$$\prod_{n=0}^{\infty} (1 - x^{4(n+1)})(1 - x^{4n+3})(1 - x^{4n+1})$$

$$= \prod_{n=0}^{\infty} (1 - x^{16(n+1)})(1 + x^{16n+10})(1 + x^{16n+6})$$

$$- \prod_{n=0}^{\infty} (1 - x^{16(n+1)})(1 + x^{16n+2})(1 + x^{16n+14}).$$

If we now divide by the left-hand side of the equation, we immediately have the following. ·

Lemma 2. *Let* $S = \{n, n > 0 \,|\, n$ *is odd or* $n \equiv \pm 4, \pm 6, 8 \pmod{16}\}$ *and let* $T = \{n, n > 0 \,|\, n$ *is odd or* $n \equiv \pm 2, \pm 4, 8 \pmod{16}\}$. *Denote by* $P(\pm 6, n, S)$ *the number of partitions of* n *from* S *disallowing repetitions of numbers congruent to* $\pm 6 \bmod 16$. *and by* $P(\pm 2, n, T)$ *the analogous quantity for* T. *Then* $P(\pm 6, n, S) = P(\pm 2, n - 1, T)$.

The above is simply the combinatorial interpertation of the identity. The interesting fact, though, is that we can algebraically convert this identity to another one in which we no longer will have any restriction on the partitions.

It is clear that

$$\prod_{n=0}^{\infty} (1 - x^{16(n+1)})(1 + x^{16n+10})(1 + x^{16n+6})$$

$$= \prod_{n=0}^{\infty} (1 - x^{16(n+1)}) \left(\frac{1 - x^{32n+20}}{1 - x^{16n+10}} \right) \left(\frac{1 - x^{32n+12}}{1 - x^{16n-6}} \right)$$

and that

$$\prod_{n=0}^{\infty} (1 - x^{16(n+1)})(1 + x^{16n+2})(1 + x^{16n+14})$$

$$= \prod_{n=0}^{\infty} (1 - x^{16(n+1)}) \left(\frac{1 - x^{32n+4}}{1 - x^{16n+2}} \right) \left(\frac{1 - x^{32n+28}}{1 - x^{16n+14}} \right),$$

and hence using these new expressions on the right-hand side of the identity and once again dividing by the left-hand side, all the numerators cancel, and we obtain the following.

Theorem 9. *Let* $S = \{n, n > 0 \mid n$ *is odd or* $n \equiv \pm 4, \pm 6, \pm 8, \pm 10$ (mod 32)$\}$ *and let* $T = \{n, n > 0 \mid n$ *is odd or* $n \equiv \pm 2, \pm 8, \pm 12, \pm 14$ (mod 32)$\}$. *Then* $P(n, S) = P(n - 1, T)$.

In a totally similar fashion the case $N = 5, l = 3$ yields the following.

Theorem 10. *Let* $S = \{n, n > 0 \mid n \equiv \pm 1$ (mod 5)$, n \equiv \pm 7$ (mod 20)$\}$ *and let* $T = \{n, n > 0 \mid n \equiv \pm 1$ (mod 5)$, n \equiv \pm 3$ (mod 20)$\}$. *Then* $P(\pm 7, n, S) = P(\pm 3, n - 1, T)$. *Alternatively, if we let*

$$S = \{n, n > 0 \mid n \equiv \pm 5, \pm 10, \pm 15, \pm 7, \pm 13, \pm 1, \pm 4, \pm 6, \pm 9, \pm 11, \pm 16, \pm 19$$

$$\text{(mod 40)}\}$$

and let

$$T = \{n, n > 0 \mid n \equiv \pm 5, \pm 10, \pm 15, \pm 3, \pm 17, \pm 1, \pm 4, \pm 9, \pm 11, \pm 14, \pm 16, \pm 19$$

$$\text{(mod 40)}\},$$

then $P(n, S) = P(n - 1, T)$.

Proof. The proof of the first statement is the combinatorial content of the identity in the infinite product version. The second statement as in the case of the previous theorem is algebraically derived from the first statement as in the proof of the previous theorem. $\square$

The case $N = 5, l = 1$ would give the last example in Andrews' paper [10]. We leave that for the reader.

The point of these examples is twofold. First, the combinatorial results are shown to follow from the most elementary properties of theta functions, and second, there is always a combinatorial identity that follows. It generally is a partition identity that places restrictions on the partitions. It can also involve partitions that involve colored numbers to distinguish among them. The reader is invited to compute the identity with $N = 3, l = 1$ for such an example. Calculations seem to indicate that the three examples found by Andrews are in fact the only ones that follow from this particular identity. Garvan's listing of many possible examples, however, shows that there are probably some other 3-term theta constant identities that are the source of these examples. We have not yet found the secret.

It is, however, possible to ask for a different sort of generalization. We can ask for three sets S, T, U such that

$$P(n, s) = P(n - a, T) + P(n - b, U).$$

In what follows we shall show that equation (4) with $N = 3$ gives some solutions to this problem in the same way that the case $N = 2$ gave solutions to the previous problem.

We begin by writing the $N = 3$ version of equation (4):

$$\theta \begin{bmatrix} \frac{l}{k} \\ 1 \end{bmatrix} (0, k\tau) = \theta \begin{bmatrix} \frac{l}{3k} \\ 1 \end{bmatrix} (0, 9k\tau) + \theta \begin{bmatrix} \frac{l+2k}{3} \\ 1 \end{bmatrix} (0, 9k\tau) + \theta \begin{bmatrix} \frac{l-2k}{3} \\ 1 \end{bmatrix} (0, 9k\tau).$$

The Jacobi triple product formula gives

$$\prod_{n=0}^{\infty} (1 - x^{2k(n+1)})(1 - x^{2kn+k+l})(1 - x^{2kn+k-l})$$

$$= \prod_{n=0}^{\infty} (1 - x^{18k(n+1)})(1 - x^{18kn+9k+3l})(1 - x^{18kn+9k-3l})$$

$$- x^{k+l} \prod_{n=0}^{\infty} (1 - x^{18k(n+1)})(1 - x^{18kn+15k+3l})(1 - x^{18kn+3k-3l})$$

$$- x^{k-l} \prod_{n=0}^{\infty} (1 - x^{18k(n+1)})(1 - x^{18kn+3k+3l})(1 - x^{18kn+15k-3l}).$$

We now consider the case $k = 3$ and $l = 1$. In this case we also can use the variable $y = x^2$. We obtain

$$\prod_{n=0}^{\infty} (1 - x^{3n+3})(1 - x^{3n+2})(1 - x^{3n+1})$$

$$= \prod_{n=0}^{\infty} (1 - x^{27(n+1)})(1 - x^{27n+15})(1 - x^{27n+12})$$

$$- x^2 \prod_{n=0}^{\infty} (1 - x^{27(n+1)})(1 - x^{27n+24})(1 - x^{27n+3})$$

$$- x \prod_{n=0}^{\infty} (1 - x^{27(n+1)})(1 - x^{27n+6})(1 - x^{27n+21}).$$

This identity gives the following result.

Theorem 11. *Let*

$$S = \{n, n > 0 \mid n \not\equiv \pm 12 \pmod{27}\},$$

$$T = \{n, n > 0 \mid n \not\equiv \pm 3 \pmod{27}\},$$

$$U = \{n, n > 0 \mid n \not\equiv \pm 6 \pmod{27}\}.$$

Then

$$P(n, S) = P(n - 2, T) + P(n - 1, U).$$

Our final example will be $k = 2$ and $l = 1$. The identity in this case is

$$\prod_{n=0}^{\infty}(1 - x^{4(n+1)})(1 - x^{4n+3})(1 - x^{4n+1})$$

$$= \prod_{n=0}^{\infty}(1 - x^{36(n+1)})(1 - x^{36n+21})(1 - x^{36n+15})$$

$$- x^3 \prod_{n=0}^{\infty}(1 - x^{36(n+1)})(1 - x^{36n+33})(1 - x^{36n+3})$$

$$- x \prod_{n=0}^{\infty}(1 - x^{36(n+1)})(1 - x^{36n+9})(1 - x^{36n+27}).$$

This gives rise to the following partition identity.

Theorem 12. *Let*

$S = \{n, n > 0 \mid n$ *is odd*$, n \not\equiv \pm 15 \pmod{36}, n \equiv \pm 4, \pm 8, \pm 12, \pm 16, 18$
$\pmod{36}\}$,

$T = \{n, n > 0 \mid n$ *is odd*$, n \not\equiv \pm 3 \pmod{36}, n \equiv \pm 4, \pm 8, \pm 12, \pm 16, 18$
$\pmod{36}\}$,

$U = \{n, n > 0 \mid n$ *is odd*$, n \not\equiv \pm 9 \pmod{36}, n \equiv \pm 4, \pm 8, \pm 12, \pm 16, 18$
$\pmod{36}\}$.

Then

$$P(n, S) = P(n - 3, T) + P(n - 1, U).$$

10 Complex Analysis

This paper up to this point has tried to show how the theory of theta functions is very useful as a guide to problems in combinatorial number theory. In this final section we shall try to demonstrate the versatility of the theta function by showing how it can also be useful in complex function theory.

High points of an elementary course in complex analysis are Picard's theorem and the Riemann mapping theorem.

Theorem 13 (Picard). *Let f be an entire function (analytic on $\mathbb{C}$) that omits two values. Then f is constant.*

Theorem 14 (Riemann mapping theorem). *Let D be a simply connected domain in $\mathbb{C}$ with at least two boundary points. Then there is an analytic homeomorphism of D onto the unit disk.*

As an example of the latter we let D be the interior of a rectangle (with sides parallel to the real and imaginary axes). Then there is an analytic homeomorphism of D onto the unit disk.

In what follows we shall show how Picard's theorem follows from a simple application of what we have called the τ-theory, in particular from Theorem 1.

Proof. Consider $g(\tau) = \theta^8 \begin{bmatrix} 0 \\ 0 \end{bmatrix} (0, \tau)/\theta^8 \begin{bmatrix} 1 \\ 0 \end{bmatrix} (0, \tau)$ and use the τ-theory to show that it is automorphic with respect to the group $\Gamma(2)$. Recall that $\Gamma(2)$ is the subgroup of $\mathrm{PSL}(2, \mathbb{Z})$ generated by $\gamma_1(\tau) = \tau + 2$ and $\gamma_2(\tau) = \frac{\tau}{2\tau+1}$. One now shows that $g(\tau)$ has a meromorphic extension to the punctures of $H/\Gamma(2)$ and that the function defined on the compactification of $H/\Gamma(2)$ has a double zero at one puncture, a double pole at the other puncture, and is finite at the third. Taking a square root thus gives us an analytic homeomorphism of the compactification onto the sphere and an analytic homeomorphism of $H/\Gamma(2)$ onto the sphere less three points, in fact, in our case the points

$$0, \quad 1, \quad \infty.$$

This allows us to define a local inverse to $\sqrt{g(\tau)}$. Call this map h.

Consider now an entire function f that omits the values $0, 1, \infty$. The function $h(f(\tau))$ is well defined on all of $\mathbb{C}$, since $\mathbb{C}$ is simply connected and it is locally well defined. It is a map of $\mathbb{C}$ into H and thus must be constant by Liouville's theorem. This is a contradiction, so no such f can exist. $\square$

In the above proof we have omitted the computations that demonstrate that the map indeed has a meromorphic extension, but the reader is capable of making this computation. We are indeed using the monodromy theorem but hope that the reader will agree that the proof modulo these points is indeed elementary. I also point out that one of the first functions one would study after learning of the τ-theory is precisely the function $g(\tau)$ given above.

We now turn our attention to giving an explicit conformal map of the rectangle with vertices at $0, 1, 1 + it, it$ $(t > 0)$ onto the open unit disk. Define

$$H(\zeta) = \frac{\theta \begin{bmatrix} 1 \\ 0 \end{bmatrix} (0, it)\theta \begin{bmatrix} 1 \\ 0 \end{bmatrix} (\zeta, it)}{\theta \begin{bmatrix} 0 \\ 0 \end{bmatrix} (0, it)\theta \begin{bmatrix} 0 \\ 0 \end{bmatrix} (\zeta, it)}$$

and

$$J(\zeta) = \frac{\theta \begin{bmatrix} 0 \\ 1 \end{bmatrix} (0, it)\theta \begin{bmatrix} 0 \\ 1 \end{bmatrix} (\zeta, it)}{\theta \begin{bmatrix} 0 \\ 0 \end{bmatrix} (0, it)\theta \begin{bmatrix} 0 \\ 0 \end{bmatrix} (\zeta, it)}.$$

Theorem 15. *The function $H + iJ$ is a conformal map of the rectangle with vertices at $0, 1, 1 + it, it$ onto the interior of the unit disk. The function $H - iJ$ is a conformal map of the rectangle with vertices at $0, 1, 1 + it, it$ onto the exterior of the unit disk.*

Proof. The proof is a sequence of observations:

(1) Consider the larger rectangle with vertices $0, 2, 2 + 2it, 2it$.
(2) The functions $H + iJ, H - iJ$ are elliptic functions with periods $2, 2it$, and this large rectangle is a fundamental parallelogram for them.
(3) The functions have at most four possible poles in this rectangle, the points

$$\frac{1 + it}{2}, \quad \frac{3 + it}{2}, \quad \frac{1 + 3it}{2}, \quad \frac{3 + 3it}{2}.$$

A computation shows that there are only two poles for each of the functions.
(4) It thus follows that $H + iJ$ $(H - iJ)$ has two poles and assumes every value twice.
(5) The theta function properties show that

$$H + iJ(\zeta + 1) = (-H + iJ)(\zeta), H + iJ(\zeta + it) = (H - iJ)(\zeta).$$

(6) The functions H, J separately are real on the lines

$$\text{Re}(\zeta) = 0, \text{Re}(\zeta) = 1, \text{Re}(\zeta) = 2,$$

$$\text{Im}(\zeta) = 0, \text{Im}(\zeta) = t, \text{Im}(\zeta) = 2t.$$

(7) The previous observation together with equation (7) gives that the lines are mapped to the unit circle. The above observations prove the theorem constructively. $\square$

References

1. H.M. Farkas, *On an Arithmetical Function*, Ramanujan Journal vol. 8 no. 3 pp. 309–315 (2004).
2. H.M. Farkas, Y. Godin, *Logrithmic Derivatives of Theta Functions*, Israel Jnl. of Math. vol. 148, pp. 253–265 (2005).
3. H.M. Farkas *Sums of Squares and Triangular Numbers* Online Journal of Analytic Combinatorics vol. 1 (2006).
4. H.M. Farkas, I. Kra, *Riemann Surfaces*, Springer-Verlag (1980).
5. H.M. Farkas, I. Kra, *Theta Constants Riemann Surfaces and the Modular Group*, AMS Grad Studies in Math, vol. 37 (2001).
6. H.E. Rauch, H.M. Farkas, *Theta Functions with Applications to Riemann Surfaces*, Williams and Wilkins, Balt. Md. (1974).
7. C.H. Clemmens, *A Scrapbook of Complex Curve Theory*, Plenum Press (1980).
8. H.M. Farkas, *On an Arithmetical Function II*, Contemp. Math. 382, Complex Analysis and Dynamical Systems II (2005).
9. R.D.M. Accola, *Theta Functions and Abelian Automorphism Groups*, Lecture Notes in Math, Springer-Verlag (1975).
10. G.E. Andrews, *Further Problems on Partitions*, American Math Monthly, May 1987, pp. 437–439.
11. F.G. Garvan, *Shifted and shiftless partition identities*, In: Number theory for the millennium, II (Urbana, IL, 2000), pp. 75–92, A.K. Peters, Natick, MA, 2002.

5

Inverse Problems for Representation Functions in Additive Number Theory

Melvyn B. Nathanson

Department of Mathematics, Lehman College (CUNY), Bronx, New York 10468
melvyn.nathanson@lehman.cuny.edu
School of Mathematics, Institute for Advanced Study, Princeton, NJ 08540
melvyn@ias.edu

Summary. For every positive integer h, the representation function of order h associated to a subset A of the integers or, more generally, of any group or semigroup $\mathbf{X}$, counts the number of ways an element of $\mathbf{X}$ can be written as the sum (or product, if $\mathbf{X}$ is nonabelian) of h not necessarily distinct elements of $\mathbf{X}$. The direct problem for representation functions in additive number theory begins with a subset A of $\mathbf{X}$ and seeks to understand its representation functions. The inverse problem for representation functions starts with a function $f : \mathbf{X} \to \mathbf{N}_0 \cup \{\infty\}$ and asks whether there is a set A whose representation function is f, and, if the answer is yes, to classify all such sets. This paper is a survey of recent progress on the inverse representation problem.

Key words: Additive bases, sumsets, representation functions, additive inverse problems asymptotic density, Erdős–Turán conjecture, Sidon sets, $B_h[g]$ sequences.

2000 *Mathematics Subject Classifications*: 11B13, 11B34, 11B05, 11A07, 11A41

1 Asymptotic Density

Let $\mathbf{N}, \mathbf{N}_0$, and $\mathbf{Z}$ denote, respectively, the sets of positive integers, nonnegative integers, and all integers.

For any set A of integers, we define the *counting function* $A(y, x)$ of A by

$$A(y, x) = \sum_{\substack{a \in A \\ y \leq a \leq x}} 1$$

for all real numbers x and y. We define the *upper asymptotic density*

$$d_U(A) = \limsup_{x \to \infty} \frac{A(-x, x)}{2x + 1}$$

The work of M.B.N. was supported in part by grants from the NSA Mathematical Sciences Program and the PSC-CUNY Research Award Program.

K. Alladi (ed.), *Surveys in Number Theory*, DOI: 10.1007/978-0-387-78510-3_5,

and the *lower asymptotic density*

$$d_L(A) = \liminf_{x \to \infty} \frac{A(-x, x)}{2x + 1}.$$

The set A has *asymptotic density* $d(A) = \alpha$ if $d_U(A) = d_L(A) = \alpha$ or, equivalently, if

$$d(A) = \lim_{x \to \infty} \frac{A(-x, x)}{2x + 1} = \alpha.$$

Let $B = \mathbf{Z} \setminus A$. Then $d_U(A) = \alpha$ if and only if $d_L(A) = 1 - \alpha$. If S and W are sets of integers, then the set W has *relative upper asymptotic density* $d_U(W, S) = \alpha$ with respect to S if

$$\limsup_{x \to \infty} \frac{(W \cap S)(-x, x)}{S(-x, x)} = \alpha.$$

Relative lower asymptotic density $d_L(W, S)$ and relative density $d(W, S)$ are defined similarly. In particular, if $S = \mathbf{N}$ and A is a set of positive integers, then A has relative asymptotic density α with respect to $\mathbf{N}$ if

$$\lim_{x \to \infty} \frac{A(-x, x)}{\mathbf{N}(-x, x)} = \lim_{x \to \infty} \frac{A(1, x)}{[x]} = \alpha.$$

2 Sumsets and Bases

Let A_1 and A_2 be subsets of an additive abelian semigroup $\mathbf{X}$. We define the *sumset*

$$A_1 + A_2 = \{a_1 + a_2 : a_1 \in A_1 \text{ and } a_2 \in A_2\}.$$

For every positive integer $h \geq 3$, if $A_1, A_2, \ldots, A_h$ are subsets of $\mathbf{X}$, then we define the sumset $A_1 + \cdots + A_{h-1} + A_h$ inductively by

$$A_1 + \cdots + A_{h-1} + A_h = (A_1 + \cdots + A_{h-1}) + A_h.$$

If $A = A_i$ for $i = 1, \ldots, h$, then we write

$$hA = \underbrace{A + \cdots + A}_{h \text{ times}}.$$

The set hA is called the *h-fold sumset* of A.

We define $0A = \{0\}$.

If $A \subseteq \mathbf{X}$ and $x \in \mathbf{X}$, we define the *shift* $A + x = A + \{x\}$.

A central concept in additive number theory is *basis*. Let S be a subset of $\mathbf{X}$. The set A is called

(1) a *basis of order h for S* if $S \subseteq hA$, that is, if every element of S can be represented as the sum of h not necessarily distinct elements of A,

(2) an *asymptotic basis of order h for S* if $S \setminus hA$ is finite, that is, if all but finitely many elements of S can be represented as the sum of h not necessarily distinct elements of A.

For example, the nonnegative cubes are a basis of order 9 for $\mathbf{N}_0$ (Wieferich's theorem), an asymptotic basis of order 7 for $\mathbf{N}_0$ (Linnik's theorem), and a basis of order 4 for almost all $\mathbf{N}_0$ (Davenport's theorem). A large part of classical additive number theory is the study of how special sets of integers (for example, the kth powers, polygonal numbers, and primes) are bases for the nonnegative integers (cf. Nathanson [13]).

Our definition of basis is weak in the sense that if $\mathbf{X}$ is an abelian semigroup with additive identity 0, then every subset of $\mathbf{X}$ has a basis of order h for all $h \geq 1$. The reason is that $0 \in \mathbf{X}$ implies that $h\mathbf{X} = \mathbf{X}$ for every positive integer h, and so, if S is any subset of $\mathbf{X}$, then $S \subseteq h\mathbf{X}$. We shall call the subset A of $\mathbf{X}$

(1) an *exact basis of order h for S* if $S = hA$, that is, if the elements of S are precisely the elements of $\mathbf{X}$ that can be represented as the sum of h not necessarily distinct elements of A,
(2) an *exact asymptotic basis of order h for S* if $hA \subseteq S$ and $S \setminus hA$ is finite.

In additive subsemigroups of the integers, the set A is a *basis of order h for almost all S* if $S \setminus hA$ has relative asymptotic density zero with respect to S, and an *exact basis of order h for almost all S* if $hA \subseteq S$ and $S \setminus hA$ has relative asymptotic density zero with respect to S.

3 Direct and Inverse Problems for Sumsets

Let $\mathbf{X}$ be an additive abelian semigroup. Given subsets $A_1, \ldots, A_h$ of $\mathbf{X}$, a *direct problem* in additive number theory is to describe the sumset $A_1 + \cdots + A_h$. In particular, for any $A \subseteq \mathbf{X}$, the direct problem is to describe the h-fold sumsets hA for all $h \geq 2$. If $\mathbf{X}$ contains an additive identity 0 and if $0 \in A \subseteq \mathbf{X}$, then we obtain an increasing sequence of sumsets

$$A \subseteq 2A \subseteq \cdots \subseteq hA \subseteq (h+1)A \subseteq \cdots . \tag{1}$$

An important open problem is to describe the evolution of structure in the sequence $\{hA\}_{h=1}^{\infty}$. For example, let $\mathbf{X} = \mathbf{N}_0$ be the additive semigroup of nonnegative integers. Let A be a set of nonnegative integers such that $d_L(h_0 A) > 0$ for some positive integer h_0. By translation and contraction, we can assume that $0 \in A$ and $\gcd(A) = 1$. Then the sequence (1) eventually stabilizes as a cofinite subset of $\mathbf{N}_0$, that is, there exists an integer $h_1 \geq h_0$ such that $h_1 A$ contains all sufficiently large integers and $hA = h_1 A$ for all $h \geq h_1$ (Nash–Nathanson [10]). However, if $d_L(hA) = 0$ for all positive integers h, then the structure of the sumsets hA is mysterious. It must happen that very regular infinite configurations of integers develop in the sumsets, but nothing is known about them.

The simplest *inverse problem for sumsets* is:

> What sets are sumsets?

This can be called the *sumset recognition problem*: Given a subset S of the abelian semigroup $\mathbf{X}$ and an integer $h \geq 2$, do there exist subsets $A_1, \ldots, A_h \subseteq \mathbf{X}$ such that

$$S = A_1 + \cdots + A_h?$$

Similarly, we have *basis recognition problems*. Let S be a subset of the abelian semigroup $\mathbf{X}$, and let $h \geq 2$. Does there exist an exact basis of order h for S, that is, a set $A \subseteq \mathbf{X}$ such that $hA = S$? If S does have an exact basis, describe the set

$$\mathcal{E}_h(S) = \{A \subseteq \mathbf{X} : S = hA\}.$$

More generally, do there exist exact asymptotic bases of order h for S? If so, describe the set

$$\mathcal{E}_h^{\text{asy}}(S) = \{A \subseteq \mathbf{X} : hA \subseteq S \text{ and } \operatorname{card}(S \setminus hA) < \infty\}.$$

4 Representation Functions of Semigroups

Let $\mathbf{X}$ be an abelian semigroup, written additively. For $A \subseteq \mathbf{X}$, let A^h denote the set of all h-tuples of A. Two h-tuples $(a_1, \ldots, a_h) \in \mathbf{X}^h$ and $(a_1', \ldots, a_h') \in \mathbf{X}^h$ are equivalent if there is a permutation $\tau : \{1, \ldots, h\} \to \{1, \ldots, h\}$ such that $a_{\tau(i)} = a_i'$ for $i = 1, \ldots, h$. If $\mathbf{X}$ is the semigroup of integers or nonnegative integers (or if $\mathbf{X}$ is any totally ordered set), then every equivalence class contains a unique h-tuple $(a_1, \ldots, a_h)$ such that $a_i \leq a_{i+1}$ for $i = 1, \ldots, h - 1$.

Let $A_1, \ldots, A_h$ be subsets of $\mathbf{X}$ and let x be an element of $\mathbf{X}$. We define the *ordered representation function*

$$R_{A_1, \ldots, A_h}(x) = \operatorname{card}\left(\{(a_1, \ldots, a_h) \in A_1 \times \cdots \times A_h : a_1 + \cdots + a_h = x\}\right).$$

If $A_i = A$ for $i = 1, \ldots, h$, then we write

$$R_{A,h}(x) = \operatorname{card}(\{(a_1, \ldots, a_h) \in A^h : a_1 + \cdots + a_h = x\}).$$

Two other representation functions arise often and naturally in additive number theory. The *unordered representation function* $r_{A,h}(x)$ counts the number of equivalence classes of h-tuples $(a_1, \ldots, a_h)$ such that $a_1 + \cdots + a_h = x$. The *unordered restricted representation function*[1] $\hat{r}_{A,h}(x)$ counts the number of equivalence classes of h-tuples $(a_1, \ldots, a_h)$ of distinct elements of $\mathbf{X}$ such that $a_1 + \cdots + a_h = x$.

---

[1] We could also introduce an *ordered restricted representation function* $\hat{R}_{A,h}(x)$ that counts the number of h-tuples $(a_1, \ldots, a_h)$ of distinct elements of $\mathbf{X}$ such that $a_1 + \cdots + a_h = x$. This is unnecessary, however, because $\hat{R}_{A,h}(x) = h!\hat{r}_{A,h}(x)$ for all $x \in \mathbf{X}$. The relation between the ordered and unordered representation functions $R_{A,h}(x)$ and $r_{A,h}(x)$ is more complex.

If **X** is a subsemigroup of the integers or of any totally ordered semigroup, then

$$r_{A,h}(x) = \text{card}(\{(a_1, \ldots, a_h) \in A^h : a_1 \leq \cdots \leq a_h \text{ and } a_1 + \cdots + a_h = x\})$$

and

$$\hat{r}_{A,h}(x) = \text{card}(\{(a_1, \ldots, a_h) \in A^h : a_1 < \cdots < a_h \text{ and } a_1 + \cdots + a_h = x\}).$$

5 Direct and Inverse Problems for Representation Functions

A fundamental *direct problem* in additive number theory is to describe the representation functions of finite and infinite subsets of the integers and of other abelian semigroups. For example, if $\mathbf{X} = \mathbf{N}_0 = A$, then $R_{A,2}(n) = n + 1$ and $r_{A,2}(n) = [(n + 2)/2]$ for all $n \in \mathbf{N}_0$. If $\mathbf{X} = \mathbf{Z} = A$, then $R_{A,2}(n) = r_{A,2}(n) = \infty$ for all $n \in \mathbf{Z}$. More generally, we ask, given a semigroup **X**, a family $\mathcal{A}$ of subsets of **X**, and a positive integer h, what properties are shared by all of the representation functions associated with sets $A \in \mathcal{A}$? These are direct problems.

The simplest *inverse problem for representation functions* is:

> ### What functions are representation functions?

More precisely, if $\mathcal{A}$ is a family of subsets of a semigroup **X** and if h is a positive integer, let $\mathcal{R}_h^{\text{ord}}(\mathcal{A})$ and $\mathcal{R}_h^{\text{unord}}(\mathcal{A})$ denote, respectively, the sets of ordered and unordered representation functions of order h associated with sets $A \in \mathcal{A}$, that is,

$$\mathcal{R}_h^{\text{ord}}(\mathcal{A}) = \{R_{A,h} : A \in \mathcal{A}\}$$

and

$$\mathcal{R}_h^{\text{unord}}(\mathcal{A}) = \{r_{A,h} : A \in \mathcal{A}\}.$$

The inverse problem is to determine whether a given function f is a representation function, and, if so, to describe all sets $A \in \mathcal{A}$ such that $R_{A,h} = f$ or $r_{A,h} = f$. This is particularly interesting when $\mathcal{A}$ is the set of bases or asymptotic bases of order h for **X**.

There is an important difference between the representation functions of asymptotic bases for the integers and the nonnegative integers. If f is the unordered representation function of an asymptotic basis for a semigroup **X**, then the set $f^{-1}(0)$ is finite. If $\mathbf{X} = \mathbf{Z}$, then a fundamental theorem in additive number theory (Theorem 7) states that for every $h \geq 2$ and for every function $f : \mathbf{Z} \to \mathbf{N}_0 \cup \{\infty\}$ with $\text{card}(f^{-1}(0)) < \infty$, there exists a set A such that $r_{A,h}(n) = f(n)$ for all $n \in \mathbf{Z}$. Equivalently, if $\mathcal{A}$ is the set of all asymptotic bases of order h for **Z**, then

$$\mathcal{R}_h^{\text{unord}}(\mathcal{A}) = \{f : \mathbf{Z} \to \mathbf{N}_0 \cup \{\infty\} : \text{card}(f^{-1}(0)) < \infty\}.$$

For the semigroup of nonnegative integers, however, it is false that every function $f : \mathbf{N}_0 \to \mathbf{N}_0$ with only finitely many zeros is the unordered representation function for an asymptotic basis of order h. Indeed, very little is known about representation functions of asymptotic bases of finite order for $\mathbf{N}_0$.

6 Representation Functions for Sets of Nonnegative Integers

If A is a set of nonnegative integers, then for every positive integer h the number of representations of an integer as the sum of h elements of A is finite. We introduce the following three sets of arithmetic functions:

$$\mathcal{F}(\mathbf{N}_0) = \{f : \mathbf{N}_0 \to \mathbf{N}_0\},$$

$$\mathcal{F}_\infty(\mathbf{N}_0) = \{f : \mathbf{N}_0 \to \mathbf{N}_0 : f^{-1}(0) \text{ is a set of density } 0\},$$

and

$$\mathcal{F}_0(\mathbf{N}_0) = \{f : \mathbf{N}_0 \to \mathbf{N}_0 : f^{-1}(0) \text{ is a finite set}\}.$$

Then

$$\mathcal{F}_0(\mathbf{N}_0) \subset \mathcal{F}_\infty(\mathbf{N}_0) \subset \mathcal{F}(\mathbf{N}_0).$$

For $h \geq 2$, the set $\mathcal{F}_0(\mathbf{N}_0)$ contains the representation functions of all bases and asymptotic bases of order h for $\mathbf{N}_0$, and the set $\mathcal{F}_\infty(\mathbf{N}_0)$ contains the representation functions of all bases of order h for almost all $\mathbf{N}_0$.

Problem 1. Let $h \geq 2$. Find necessary and sufficient conditions for a function in $\mathcal{F}_0$ to be the representation function for an asymptotic basis of order h for $\mathbf{N}_0$.

Problem 2. Let $h \geq 2$. Find necessary and sufficient conditions for a function in $\mathcal{F}_\infty$ to be the representation function for a basis of order h for almost all $\mathbf{N}_0$.

Problem 3. Let $h \geq 2$. Find necessary and sufficient conditions for a function in $\mathcal{F}$ to be the representation function for a subset of $\mathbf{N}_0$.

We can also count the number of representations of a nonnegative integer as the sum of a bounded number of elements of a set that contains both nonnegative and negative integers.

Problem 4. Let $h \geq 2$. Find necessary and sufficient conditions for a function in $\mathcal{F}$ to be the representation function for the nonnegative integers in the h-fold sumset of a subset of $\mathbf{Z}$.

We can express the ordered and unordered representation functions of a set of nonnegative integers in terms of generating functions. Define the *generating function* for the set A of nonnegative integers as the power series

$$G_A(z) = \sum_{a \in A} z^a.$$

This can be used both as a formal power series and as an analytic function that converges for $|z| < 1$. We have the identities

$$\sum_{n=0}^{\infty} r_{A,2}(n)z^n = \frac{1}{2}\left(G_A^2(z) + G_A(z^2)\right),$$

$$\sum_{n=0}^{\infty} \hat{r}_{A,2}(n)z^n = \frac{1}{2}\left(G_A^2(z) - G_A(z^2)\right),$$

and, for all $h \geq 1$,

$$\sum_{n=0}^{\infty} R_{A,h}(n)z^n = G_A^h(z).$$

If A is a set of integers, then the ordered representation function $R_{A,2}(n)$ is odd if and only if n is even and $n/2 \in A$. It follows that $R_{A,2}(n)$ is eventually constant if and only if A is finite. Moreover, the ordered representation function $R_{A,2}$ uniquely determines the set A. Thus, for every function $f \in \mathcal{F}(\mathbf{N}_0)$, there exists at most one set A such that $R_{A,2} = f$. Theorem 3 generalizes this observation to all $h \geq 2$.

It is also true that the unordered representation function $r_{A,2}(n)$ for a set A of nonnegative integers is eventually constant only if A is finite.

Theorem 1 (Dirac [4]). *If A is an infinite set of nonnegative integers, then the representation function $r_{A,2}(n)$ is not eventually constant.*

Proof. Let A be an infinite set of nonnegative integers such that $r_{A,2}(n) = c$ for all $n \geq n_0$. Since A is infinite, we have $r_{A,2}(2a) \geq 1$ for all $a \in A$, and so $c \geq 1$. There is a polynomial $P(z)$ such that

$$\frac{1}{2}(G_A^2(z) + G_A(z^2)) = \sum_{n=0}^{\infty} r_{A,2}(n)z^n$$

$$= \sum_{n=0}^{n_0-1} r_{A,2}(n)z^n + \sum_{n=n_0}^{\infty} cz^n$$

$$= \frac{P(z)}{1-z}.$$

Let $0 < x < 1$ and $z = -x$. Then $G_A(z) = G_A(-x)$ is real, and so $G_A^2(z) \geq 0$ and

$$\frac{2P(-x)}{1+x} = G_A^2(-x) + G_A(x^2) \geq G_A(x^2).$$

Taking the limit as $x \to 1^-$, we see that the left side of this equality converges to $P(-1)$, but the right side diverges to infinity. This is impossible, and so the representation function $r_{A,2}(n)$ cannot be eventually constant. $\qquad\square$

Dirac's theorem is a special case of a famous unsolved problem in additive number theory. Erdős and Turán [5] conjectured that if A is an asymptotic basis of order 2 for the nonnegative integers, then $\lim \sup_{n \to \infty} r_{A,2}(n) = \infty$. This conjecture is itself only a small part of the problem of characterizing the representation functions of additive bases of finite order for $\mathbf{N}_0$. It is interesting to note that the modular analogue of the Erdős–Turán conjecture is false.

Theorem 2 (Tang–Chen [22]). *There is an integer m_0 such that for every $m \geq m_0$, there is a set $A_m \subseteq \mathbf{Z}/m\mathbf{Z}$ such that A_m is a basis of order 2 for $\mathbf{Z}/m\mathbf{Z}$ and $r_{A_m,2}(x) \leq 768$ for all $x \in \mathbf{Z}/m\mathbf{Z}$.*

It is also interesting that the multiplicative Erdős and Turán conjecture is true. If A is a set of positive integers such that every sufficiently large positive integer is the product of two elements of A, then the number of representations of an integer n as the product of two elements of A is unbounded (Erdős [6], Nešetřil and Rödl [19], Nathanson [12]).

6.1 Ordered Representation Functions

The first inverse theorems for ordered representation functions of sets of nonnegative integers are the following.

Theorem 3 (Nathanson [11]). *Let $h \geq 2$. If A and B are sets of nonnegative integers such that $R_{A,h}(n) = R_{B,h}(n)$ for all $n \in \mathbf{N}_0$, then $A = B$.*

Proof. Since $A = \emptyset$ if and only if $B = \emptyset$, we can assume that both A and B are nonempty sets. Then the generating functions

$$G_A(z) = \sum_{a \in A} z^a \qquad \text{and} \qquad G_B(z) = \sum_{b \in B} z^b$$

are nonzero power series with nonnegative coefficients. We have

$$G_A^h(z) = \left(\sum_{a \in A} z^a \right)^h = \sum_{n=0}^{\infty} R_{A,h}(n) = \sum_{n=0}^{\infty} R_{B,h}(n) = \left(\sum_{b \in B} z^b \right)^h = G_B^h(z),$$

and so

$$0 = G_A^h(z) - G_B^h(z) = (G_A(z) - G_B(z)) \left(\sum_{i=0}^{h-1} G_A^{h-1-i}(z) G_B^i(z) \right).$$

The coefficients of the power series $\sum_{i=0}^{h-1} G_A^{h-1-i}(z) G_B^i(z)$ are nonnegative and not all zero; hence this series is nonzero, and so $G_A(z) - G_B(z) = 0$. This implies that $A = B$. $\qquad \square$

Let A^*, B^*, and T be finite sets of integers. If each residue class modulo m contains exactly the same number of elements of A^* as elements of B^*, then we write $A^* \equiv B^* \pmod{m}$. If for each integer n the number of pairs $(a, t) \in A^* \times T$ such that $a + t \equiv n \pmod{m}$ equals the number of pairs $(b, t) \in B^* \times T$ such that $b + t \equiv n \pmod{m}$, then we write

$$A^* + T \equiv B^* + T \pmod{m}.$$

Theorem 4 (Nathanson [11]). *Let A and B be sets of nonnegative integers. Then $R_{A,2}(n) = R_{B,2}(n)$ for all sufficiently large n if and only if there exist*

(i) a nonnegative integer n_0 and sets $A^*, B^* \subseteq \{0, 1, 2, \ldots, n_0\}$, and
(ii) a positive integer m and a set $T \subseteq \{0, 1, 2, \ldots, m-1\}$ with

$$A^* + T \equiv B^* + T \pmod{m}$$

such that
$$A = A^* \cup C \quad \text{and} \quad B = B^* \cup C, \tag{2}$$

where
$$C = \{c \in \mathbf{N}_0 : c > n_0 \text{ and } c \equiv t \pmod{m} \text{ for some } t \in T\}. \tag{3}$$

Proof. Let n_0 and m be integers and let A^*, B^*, and T be finite sets of integers satisfying conditions (i) and (ii). Define the sets A, B, and C by (2) and (3). Since $A^* \cap C = \emptyset$ and $B^* \cap C = \emptyset$, it follows that for every integer n we have

$$R_{A,2}(n) = R_{A^*,2}(n) + 2R_{A^*,C}(n) + R_{C,2}(n)$$

and

$$R_{B,2}(n) = R_{B^*,2}(n) + 2R_{B^*,C}(n) + R_{C,2}(n),$$

where $R_{A^*,C}(n)$ (respectively $R_{B^*,C}(n)$) is the number of ordered pairs $(a^*, c) \in A^* \times C$ (respectively $(b^*, c) \in B^* \times C$) such that $a^* + c = n$ (respectively $b^* + c = n$).

Let $n > 2n_0$. Since $\max(A^* \cup B^*) \leq n_0$, it follows that $R_{A^*,2}(n) = R_{B^*,2}(n) = 0$, and so $R_{A,2}(n) = R_{B,2}(n)$ if and only if $R_{A^*,C}(n) = R_{B^*,C}(n)$. If $a^* \in A$, then $n - a^* > 2n_0 - a^* \geq n_0$. It follows that $n - a^* \in C$ if and only if $n - a^* \equiv t \pmod{m}$ for some $t \in T$. Since $A^* + T \equiv B^* + T \pmod{m}$, it follows that

$$R_{A^*,C}(n) = \text{card}(\{(a^*, c) \in A^* \times C : a^* + c = n\})$$

$$= \sum_{t \in T} \text{card}(\{(a^*, c) \in A^* \times C : a^* + c = n \text{ and } c \equiv t \pmod{m}\})$$

$$= \sum_{t \in T} \text{card}(\{(a^*, t) \in A^* \times T : a^* + t \equiv n \pmod{m}\})$$

$$= \sum_{t \in T} \text{card}(\{(b^*, t) \in B^* \times T : b^* + t \equiv n \pmod{m}\})$$

$$= \sum_{t \in T} \text{card}(\{(b^*, c) \in B^* \times C : b^* + c = n \text{ and } c \equiv t \pmod{m}\})$$

$$= R_{B^*,C}(n).$$

Thus, the representation functions of the sets A and B eventually coincide.

Conversely, let A and B be distinct sets of integers such that $R_{A,2}(n) = R_{B,2}(n)$ for all integers $n > n_1$. Since A is finite if and only if $R_{A,2}(n) = 0$ for all sufficiently large n, it follows that the representation functions of any pair of finite sets eventually coincide, and so A is finite if and only if B is finite. Thus, we can set $A^* = A$, $B^* = B$, and $T = C = \emptyset$.

Suppose that A and B are distinct infinite sets of integers. Applying the generating functions $G_A(z) = \sum_{a \in A} z^a$ and $G_B(z) = \sum_{b \in B} z^b$, we have

$$G_A^2(z) - G_B^2(z) = \sum_{n=0}^{\infty} (R_{A,2}(n) - R_{B,2}(n)) z^n = P(z),$$

where $P(z)$ is a polynomial of degree at most n_1. The ordered representation function $R_{A,2}(n)$ (respectively $R_{B,2}(n)$) is odd if and only if n is even and $n/2 \in A$ (respectively $n/2 \in B$). It follows that the sets A and B coincide for $n > n_1/2$, and so there is a nonzero polynomial $Q(z)$ of degree at most $n_1/2$ such that

$$G_A(z) - G_B(z) = Q(z).$$

We obtain a rational function

$$G_A(z) + G_B(z) = \frac{G_A^2(z) - G_B^2(z)}{G_A(z) - G_B(z)} = \frac{P(z)}{Q(z)}.$$

Therefore, the coefficients of the power series $G_A(z) + G_B(z)$ satisfy a linear recurrence relation. For $n > n_1/2$, the coefficient of z^n in $G_A(z) + G_B(z)$ is 2 if $n \in A \cap B$ and 0 if $n \notin A \cap B$. Since a sequence defined by a linear recurrence in a finite set must be eventually periodic, it follows that there are positive integers m and n_0 and a set $T \subseteq \{0, 1, \ldots, m - 1\}$ such that for $n > n_0$, we have $n \in A \cap B$ if and only if $n \equiv t \pmod{m}$ for some $t \in T$. Let

$$C = \{c \in \mathbf{N}_0 : c > n_0 \text{ and } c \equiv t \pmod{m} \text{ for some } t \in T\}.$$

Let $A^* = A \cap [0, n_0]$ and $B^* = B \cap [0, n_0]$. Then $A^* \cap C = B^* \cap C = \emptyset$, and $A = A^* \cup C$ and $B = B^* \cup C$. For $n > 2n_0$ we have

$$2R_{A^*,C}(n) = R_{A,2}(n) - R_{C,2}(n) = R_{B,2}(n) - R_{C,2}(n) = 2R_{B^*,C}(n),$$

where as above, $R_{A^*,C}(n)$ (respectively $R_{B^*,C}(n)$) is the number of solutions of the congruence $n \equiv a + t \pmod{m}$ (respectively $n \equiv b + t \pmod{m}$) with $t \in T$ and $a \in A^*$ (respectively $b \in B^*$). Therefore, $A^* + T \equiv B^* + T \pmod{m}$, and the theorem follows.   □

Problem 5. Let $h \geq 3$. Describe all pairs of sets of nonnegative integers whose ordered representation functions of order h eventually coincide. Equivalently, classify all pairs (A, B) of sets of nonnegative integers such that $R_{A,h}(n) = R_{B,h}(n)$ for all sufficiently large integers n.

6.2 Unordered Representation Functions

Theorem 4 completely describes all pairs of sets of nonnegative integers whose ordered representation functions of order 2 eventually coincide. The analogous problem for unordered representation functions is open.

Problem 6. Describe all pairs of sets of nonnegative integers whose unordered representation functions of order 2 eventually coincide.

Problem 7. Let $h \geq 3$. Describe all pairs of sets of nonnegative integers whose unordered representation functions eventually coincide.

The behavior of unordered representation functions is more exotic than that of ordered representation functions. For example, the following beautiful result describes partitions of the nonnegative integers into disjoint sets A and B whose unordered representation functions eventually coincide.

Theorem 5 (Sándor [20]). *Let A be a set of nonnegative integers, and let $B = \mathbf{N}_0 \setminus A$. There exists a positive integer N such that $r_{A,2}(n) = r_{B,2}(n)$ for all $n \geq 2N - 1$ if and only if*

(i)
$$\mathrm{card}(A \cap [0, 2N - 1]) = N$$

(ii) for every integer $a \geq N$,
$$a \in A \text{ if and only if } 2a \notin A$$

and
$$a \in A \text{ if and only if } 2a + 1 \in A.$$

Proof. Let $\chi_A(n)$ denote the characteristic function of the set A, that is,

$$\chi_A(n) = \begin{cases} 1 & \text{if } n \in A, \\ 0 & \text{if } n \notin A. \end{cases}$$

Since $B = \mathbf{N}_0 \setminus A$, we have

$$\chi_B(n) = 1 - \chi_A(n) \qquad \text{for all } n \in \mathbf{N}_0.$$

Defining the generating functions

$$G_A(z) = \sum_{a \in A} z^a = \sum_{n=0}^{\infty} \chi_A(n) z^n$$

and

$$G_B(z) = \sum_{b \in B} z^b = \sum_{n=0}^{\infty} (1 - \chi_A(n)) z^n = \frac{1}{1 - z} - G_A(z),$$

we obtain

$$\sum_{n=0}^{\infty} r_{A,2}(n) z^n = \frac{1}{2}(G_A(z)^2 + G_A(z^2))$$

and

$$\sum_{n=0}^{\infty} r_{B,2}(n)z^n$$

$$= \frac{1}{2}(G_B(z)^2 + G_B(z^2))$$

$$= \frac{1}{2}\left(\left(\frac{1}{1-z} - G_A(z)\right)^2 + \left(\frac{1}{1-z^2} - G_A(z^2)\right)\right)$$

$$= \frac{1}{2}\left(\frac{2}{(1-z^2)(1-z)} - \frac{2G_A(z)}{1-z} + G_A(z)^2 - G_A(z^2)\right)$$

$$= \frac{1}{2}(G_A(z)^2 + G_A(z^2)) + \left(\frac{1}{(1-z^2)(1-z)} - \frac{G_A(z)}{1-z} - G_A(z^2)\right)$$

$$= \sum_{n=0}^{\infty} r_{A,2}(n)z^n + \frac{1}{1-z}\left(\frac{1}{1-z^2} - G_A(z) - (1-z)G_A(z^2)\right)$$

$$= \sum_{n=0}^{\infty} r_{A,2}(n)z^n + \frac{1}{1-z}\left(\sum_{n=0}^{\infty} z^{2n} - \sum_{n=0}^{\infty} \chi_A(n)z^n\right.$$

$$\left. - \sum_{n=0}^{\infty} \chi_A(n)z^{2n} + \sum_{n=0}^{\infty} \chi_A(n)z^{2n+1}\right)$$

$$= \sum_{n=0}^{\infty} r_{A,2}(n)z^n + \frac{1}{1-z}\left(\sum_{n=0}^{\infty} (1 - \chi_A(n) - \chi_A(2n))\, z^{2n}\right.$$

$$\left. + \sum_{n=0}^{\infty} (\chi_A(n) - \chi_A(2n+1))\, z^{2n+1}\right).$$

We define the function

$$Q(z) = \sum_{n=0}^{\infty} \left(r_{A,2}(n) - r_{B,2}(n)\right) z^n.$$

Then

$$(1-z)Q(z)$$

$$= \sum_{n=0}^{\infty} (1 - \chi_A(n) - \chi_A(2n))\, z^{2n} + \sum_{n=0}^{\infty} (\chi_A(n) - \chi_A(2n+1))\, z^{2n+1}$$

$$= \sum_{n=0}^{N-1} (1 - \chi_A(n) - \chi_A(2n)) z^{2n} + \sum_{n=0}^{N-1} (\chi_A(n) - \chi_A(2n+1)) z^{2n+1}$$

$$+ \sum_{n=N}^{\infty} (1 - \chi_A(n) - \chi_A(2n)) z^{2n} + \sum_{n=N}^{\infty} (\chi_A(n) - \chi_A(2n+1)) z^{2n+1}.$$

Let N be a positive integer. We have $r_{A,2}(n) = r_{B,2}(n)$ for all $n \geq 2N - 1$ if and only if $Q(z)$ is a polynomial of degree at most $2N - 2$. Then $(1 - z)Q(z)$ has degree at most $2N - 1$, and we have the two equations

$$(1 - z)Q(z) = \sum_{n=0}^{N-1} (1 - \chi_A(n) - \chi_A(2n)) z^{2n}$$

$$+ \sum_{n=0}^{N-1} (\chi_A(n) - \chi_A(2n+1)) z^{2n+1}$$

and

$$0 = \sum_{n=N}^{\infty} (1 - \chi_A(n) - \chi_A(2n)) z^{2n} + \sum_{n=N}^{\infty} (\chi_A(n) - \chi_A(2n+1)) z^{2n+1}.$$

If the first equation holds, then setting $z = 1$, we obtain

$$0 = \sum_{n=0}^{N-1} (1 - \chi_A(2n) - \chi_A(2n+1)) = N - \sum_{n=0}^{2N-1} \chi_A(n),$$

and so

$$\mathrm{card}(A \cap [0, 2N - 1]) = N,$$

which is condition (i). The second equation is equivalent to condition (ii). If this condition holds, then $Q(z)$ is a polynomial of degree at most $N - 2$. This completes the proof. $\square$

Problem 8. Let $\ell \geq 3$. Does there exist a partition of the nonnegative integers into disjoint sets $A_1, A_2, \ldots, A_\ell$ whose representation functions $r_{A_i,2}(n)$ for $i = 1, 2, \ldots, \ell$ eventually coincide?

7 Representation Functions for Sets of Integers

7.1 Unique Representation Bases for the Integers

Sumsets of integers are very different from sumsets of nonnegative integers. For example, the Erdős–Turán conjecture asserts that the representation function of a basis of order 2 for the nonnegative integers must be unbounded. In sharp contrast to this, there exist bases for the integers whose representation functions are bounded. Indeed, we shall construct a basis A of order 2 for $\mathbf{Z}$ whose representation function is identically equal to 1. Such sets are called *unique representation bases*.

Theorem 6 (Nathanson [14]). *Let $\varphi(x)$ be a function such that $\lim_{x \to \infty} \varphi(x) = \infty$. There exists an additive basis A for the group $\mathbf{Z}$ of integers such that*

$$r_{A,2}(n) = 1 \quad \text{for all } n \in \mathbf{Z},$$

and

$$A(-x, x) \le \varphi(x)$$

for all sufficiently large x.

Proof. We shall construct an ascending sequence of finite sets $A_1 \subseteq A_2 \subseteq A_3 \subseteq \cdots$ such that for all $k \in \mathbf{N}$ and $n \in \mathbf{Z}$,

$$|A_k| = 2k \quad \text{and} \quad r_{A_k}(n) \le 1$$

and

$$r_{A_{2k}}(n) = 1 \quad \text{if } |n| \le k.$$

It follows that the infinite set

$$A = \bigcup_{k=1}^{\infty} A_k$$

is a unique representation basis for the integers.

We construct the sets A_k by induction. Let $A_1 = \{0, 1\}$. We assume that for some $k \ge 1$ we have constructed sets

$$A_1 \subseteq A_2 \subseteq \cdots \subseteq A_k$$

such that $|A_k| = 2k$ and

$$r_{A_k}(n) \le 1 \quad \text{for all } n \in \mathbf{Z}.$$

We define the integer

$$d_k = \max\{|a| : a \in A_k\}.$$

Then

$$A_k \subseteq [-d_k, d_k]$$

and

$$2A_k \subseteq [-2d_k, 2d_k].$$

If both numbers d_k and $-d_k$ belong to the set A_k, then, since $0 \in A_1 \subseteq A_k$ and $d_k \ge 1$, we would have the following two representations of 0 in the sumset $2A_k$:

$$0 = 0 + 0 = (-d_k) + d_k.$$

This is impossible, since $r_{A_k}(0) \le 1$; hence only one of the two integers d_k and $-d_k$ belongs to the set A_k. It follows that if $d_k \notin A_k$, then

$$\{2d_k, 2d_k - 1\} \cap 2A_k = \emptyset,$$

and if $-d_k \notin A_k$, then

$$\{-2d_k, -(2d_k - 1)\} \cap 2A_k = \emptyset.$$

Select an integer b_k such that

$$b_k = \min\{|b| : b \notin 2A_k\}.$$

Then

$$1 \le b_k \le 2d_k - 1.$$

To construct the set A_{k+1}, we choose an integer c_k such that

$$c_k \ge d_k.$$

If $b_k \notin 2A_k$, let

$$A_{k+1} = A_k \cup \{b_k + 3c_k, -3c_k\}.$$

We have

$$b_k = (b_k + 3c_k) + (-3c_k) \in 2A_{k+1}.$$

If $b_k \in 2A_k$, then $-b_k \notin 2A_k$, and we let

$$A_{k+1} = A_k \cup \{-(b_k + 3c_k), 3c_k\}.$$

Again we have

$$-b_k = -(b_k + 3c_k) + 3c_k \in 2A_{k+1}.$$

Since

$$d_k < 3c_k < b_k + 3c_k,$$

it follows that $|A_{k+1}| = |A_k| + 2 = 2(k + 1)$. Moreover,

$$d_{k+1} = \max\{|a| : a \in A_{k+1}\} = b_k + 3c_k.$$

For example, since $A_1 = \{0, 1\}$ and $2A_1 = \{0, 1, 2\}$, it follows that $d_1 = b_1 = 1$. Then $b_1 \in 2A_1$, but $-1 = -b_1 \notin 2A_1$. Choose an integer $c_1 \ge 1$ and let

$$A_2 = \{-(1 + 3c_1), 0, 1, 3c_1\}.$$

Then

$$2A_2 = \{-(2 + 6c_1), -(1 + 3c_1), -3c_1, -1, 0, 1, 2, 3c_1, 1 + 3c_1, 6c_1\}$$

and $d_2 = 1 + 3c_1$ and $b_2 = 2$. Moreover, $r_{A_2}(n) = 1$ if $|n| \le 1$.

Assume that $b_k \notin 2A_k$; hence $A_{k+1} = A_k \cup \{b_k + 3c_k, -3c_k\}$. (The argument in the case $b_k \in 2A_k$ and $-b_k \notin 2A_k$ is similar.) The sumset $2A_{k+1}$ is the union of four sets:

$$2A_{k+1} = 2A_k \cup (A_k + b_k + 3c_k) \cup (A_k - 3c_k) \cup \{b_k, 2b_k + 6c_k, -6c_k\}.$$

We shall show that these sets are pairwise disjoint. If $u \in 2A_k$, then

$$-2c_k \leq -2d_k \leq u \leq 2d_k \leq 2c_k.$$

Let $a \in A_k$ and $v = a + b_k + 3c_k \in A_k + b_k + 3c_k$. The inequalities

$$-c_k \leq -d_k \leq a \leq d_k \leq c_k$$

and

$$1 \leq b_k \leq 2d_k - 1 \leq 2c_k - 1$$

imply that

$$2c_k + 1 \leq v \leq 6c_k - 1 < 2b_k + 6c_k.$$

Similarly, if $w = a - 3c_k \in A_k - 3c_k$, then

$$-6c_k < -4c_k \leq w \leq -2c_k.$$

These inequalities imply that the sets $2A_k$, $A_k + b_k + 3c_k$, $A_k - 3c_k$, and $2\{b_k + 3c_k, -3c_k\}$ are pairwise disjoint, unless $c_k = d_k$ and $-2d_k \in 2A_k \cap (A_k - 3d_k)$. If $-2d_k \in 2A_k$, then $-d_k \in A_k$. If $-2d_k \in A_k - 3d_k$, then $d_k \in A_k$. This is impossible, however, because the set A_k does not contain both integers d_k and $-d_k$.

Since the sets $A_k + b_k + 3c_k$ and $A_k - 3c_k$ are translations, it follows that

$$r_{A_{k+1}}(n) \leq 1 \qquad \text{for all integers } n.$$

Let $A = \bigcup_{k=1}^{\infty} A_k$. For all $k \geq 1$ we have $2 = b_2 \leq b_3 \leq \cdots$ and $b_k < b_{k+2}$, hence $b_{2k} \geq k + 1$. Since b_{2k} is the minimum of the absolute values of the integers that do not belong to $2A_{2k}$, it follows that

$$\{-k, -k+1, \ldots, -1, 0, 1, \ldots, k-1, k\} \subseteq 2A_{2k} \subseteq 2A$$

for all $k \geq 1$, and so A is an additive basis of order 2. In particular, $r_{A_{2k}}(n) \geq 1$ for all n such that $|n| \leq k$. If $r_{A,2}(n) \geq 2$ for some n, then $r_{A_k,2}(n) \geq 2$ for some k, which is impossible. Therefore, A is a unique representation basis for the integers.

We observe that if $x \geq 1$ and k is the unique integer such that $d_k \leq x < d_{k+1}$, then

$$A(-x, x) = A_{k+1}(-x, x)$$
$$= \begin{cases} 2k & \text{for } d_k \leq x < 3c_k, \\ 2k+1 & \text{for } 3c_k \leq x < b_k + 3c_k = d_{k+1}. \end{cases}$$

In the construction of the set A_{k+1}, the only constraint on the choice of the number c_k was that $c_k \geq d_k$. Given a function $\varphi(x)$ such that $\lim_{x \to \infty} \varphi(x) = \infty$, we shall use induction to construct a sequence of integers $\{c_k\}_{k=1}^{\infty}$ such that $A(-x, x) \leq \varphi(x)$ for all $x \geq c_1$. We begin by choosing a positive integer c_1 such that

$$\varphi(x) \geq 4 \qquad \text{for } x \geq c_1.$$

Then

$$A(-x, x) \leq 4 \leq \varphi(x) \qquad \text{for } c_1 \leq x \leq d_2.$$

Let $k \geq 2$, and suppose we have selected an integer $c_{k-1} \geq d_{k-1}$ such that

$$\varphi(x) \geq 2k \qquad \text{for } x \geq c_{k-1}$$

and

$$A(-x, x) \leq \varphi(x) \qquad \text{for } c_1 \leq x \leq d_k.$$

There exists an integer $c_k \geq d_k$ such that

$$\varphi(x) \geq 2k + 2 \qquad \text{for } x \geq c_k.$$

Then

$$A(-x, x) = 2k \leq \varphi(x) \qquad \text{for } d_k \leq x < 3c_k$$

and

$$A(-x, x) \leq 2k + 2 \leq \varphi(x) \qquad \text{for } 3c_k \leq x \leq d_{k+1};$$

hence

$$A(-x, x) \leq \varphi(x) \qquad \text{for } c_1 \leq x \leq d_{k+1}.$$

It follows that

$$A(-x, x) \leq \varphi(x) \qquad \text{for all } x \geq c_1.$$

This completes the proof. $\qquad \square$

Theorem 6 constructs arbitrarily sparse unique representation bases. If A is a unique representation basis of order 2 with counting function $A(x)$, then $A(x) \ll x^{1/2}$. We do not know how dense a unique representation basis can be.

Problem 9. Let Θ be the set of all positive numbers θ such that there exists a unique representation basis A with $A(x) \gg x^{\theta}$. Compute $\sup \Theta$.

There is work related to this problem by Chen [1] and Lee [9] for all $x \in hA_1 \setminus \{u_1\}$.

7.2 Asymptotic Bases for the Integers

Let $\mathcal{F}(\mathbf{Z})$ denote the set of all functions from $\mathbf{Z}$ into $\mathbf{N}_0 \cup \{\infty\}$. We shall consider the following two subsets of this function space: the set of functions with only finitely many zeros,

$$\mathcal{F}_0(\mathbf{Z}) = \{f \in \mathcal{F}(\mathbf{Z}) : \operatorname{card}(f^{-1}(0)) < \infty\},$$

and the set of functions that are nonzero for almost all integers n,

$$\mathcal{F}_\infty(\mathbf{Z}) = \{f \in \mathcal{F}(\mathbf{Z}) : d(f^{-1}(0)) = 0\}.$$

For every positive integer h, let $\mathcal{R}_h(\mathbf{Z})$ denote the set of all representation functions of h-fold sumsets, that is,

$$\mathcal{R}_h(\mathbf{Z}) = \{f \in \mathcal{F}(\mathbf{Z}) : f = r_{A,h} \text{ for some } A \subseteq \mathbf{Z}\}.$$

For example, $\mathcal{R}_1(\mathbf{Z}) = \{f : \mathbf{Z} \to \{0, 1\}\}$.

Let $h \geq 2$. If A is a set of integers and $a \in A$, then $r_{A,h}(ha) \geq 1$. It follows that if $f \in \mathcal{F}(\mathbf{Z})$ is a nonzero function such that $f(n) = 0$ for all $n \equiv 0 \pmod{h}$, then f is not a representation function, and so $\mathcal{F}(\mathbf{Z}) \neq \mathcal{R}_h(\mathbf{Z})$.

Problem 10. Let $h \geq 2$. Find necessary and sufficient conditions for a function $f \in \mathcal{F}(\mathbf{Z})$ to be the representation function of an h-fold sumset.

This is called the *inverse problem for representation functions in additive number theory*.

The set A is an asymptotic basis of order h for the integers if all but finitely many integers can be represented as the sum of h not necessarily distinct elements of A. Equivalently, A is an asymptotic basis of order h for $\mathbf{Z}$ if the representation function $r_{A,h}$ is an element of the function space $\mathcal{F}_0(\mathbf{Z})$. We define

$$\mathcal{R}_{h,0}(\mathbf{Z}) = \{f \in \mathcal{F}_0(\mathbf{Z}) : f = r_{A,h} \text{ for some } A \subseteq \mathbf{Z}\}.$$

Thus, $\mathcal{R}_{h,0}(\mathbf{Z})$ is the set of representation functions of asymptotic bases of order h for $\mathbf{Z}$. We shall prove the following important result: for every integer $h \geq 2$,

$$\boxed{\mathcal{R}_{h,0}(\mathbf{Z}) = \mathcal{F}_0(\mathbf{Z}).}$$

This means that *every* function $f : \mathbf{Z} \to \mathbf{N}_0 \cup \{\infty\}$ with only finitely many zeros is the representation function for some asymptotic basis of order h for the integers.

The proof will use Sidon sets. A subset A of an additive abelian semigroup $\mathbf{X}$ is called a *Sidon set of order h* if every element in the sumset hA has a unique representation (up to permutations of the summands) as a sum of h elements of $\mathbf{X}$. Equivalently, A is a Sidon set if $r_{A,h}(x) \leq 1$ for all $x \in \mathbf{X}$. Sidon sets of order h are also called B_h-sets. For example, every two-element set $\{a, b\}$ of integers (or two-element subset $\{a, b\}$ of any torsion-free abelian semigroup) is a Sidon set of order h for all positive integers h, since the h-fold sumset

$$h\{a, b\} = \{(h - i)a + ib : i = 0, 1, \ldots, h\} = \{ha + i(b - a) : i = 0, 1, \ldots, h\}$$

is simply an arithmetic progression of length $h + 1$ and difference $b - a$. Note that if the set A is a Sidon set of order h, then A is also a Sidon set of order h' for all $h' = 1, 2, \ldots, h - 1$.

The set A will be called a *generalized Sidon set of order h* if for all pairs of positive integers r, r' with $r \leq h$ and $r' \leq h$, and for all sequences $a_1, \ldots, a_r$ and $a'_1, \ldots, a'_{r'}$ of elements of A, we have

$$a_1 + \cdots + a_r = a'_1 + \cdots + a'_{r'}$$

if and only if $r = r'$ and $a'_i = a_{\sigma(i)}$ for some permutation σ of $\{1, \ldots, r\}$ and all $i = 1, \ldots, r$.

Note that if A is a Sidon set (respectively generalized Sidon set) of order h, then A is also a Sidon set (respectively generalized Sidon set) of order h' for all positive integers $h' < h$.

Lemma 1. *Let $h \geq 2$ and let c and u be integers such that $c > 2h|u|$. Then*

$$D_{c,u} = \{-c, (h - 1)c + u\}$$

is a generalized Sidon set of order h, and $u \in hD_{c,u}$. Moreover,

$$\min\left\{|x - y| : x, y \in \bigcup_{r=1}^{h} rD_{c,u} \text{ and } x \neq y\right\} \geq c/2.$$

Proof. We have

$$u = (h - 1)(-c) + ((h - 1)c + u) \in hD_{c,u}.$$

To show that $D_{c,u}$ is a generalized Sidon set, let i, j, i', j' be nonnegative integers such that

$$1 \leq i + j \leq i' + j' \leq h.$$

We define

$$\Delta = [i(-c) + j((h - 1)c + u)] - [i'(-c) + j'((h - 1)c + u)].$$

If $\Delta = 0$, then

$$(j' - j)hc = ((i' + j') - (i + j))c + (j - j')u.$$

If $j' \neq j$, then

$$hc \leq |(j' - j)hc|$$
$$= |((i' + j') - (i + j))|c + |j - j'||u|$$
$$\leq (h - 1)c + h|u|$$
$$< \left(h - \frac{1}{2}\right)c,$$

which is absurd. Therefore, $j = j'$, and so $i = i'$ and $D_{c,u}$ is a generalized Sidon set of order h.

Suppose that $\Delta \neq 0$. We must show that $|\Delta| > c/2$. If $j = j'$, then $i \neq i'$ and

$$|\Delta| = |i' - i|c \geq c.$$

If $j \neq j'$, then

$$
\begin{aligned}
|\Delta| &= \left|(j - j')hc + ((i' + j') - (i + j))|c + (j - j')|u|\right| \\
&\geq |j - j'|\,hc - \left|((i' + j') - (i + j)\right|c - \left|(j - j')u\right| \\
&\geq hc - (h - 1)c - h|u| \\
&> \frac{c}{2}.
\end{aligned}
$$

This completes the proof. □

Theorem 7 (Nathanson [15, 17]). *Let $f : \mathbf{Z} \to \mathbf{N}_0 \cup \{\infty\}$ be a function such that* card($f^{-1}(0)$) $< \infty$. *For every $h \geq 2$, there exists a set A of integers such that* $r_{A,h}(n) = f(n)$ *for all $n \in \mathbf{Z}$.*

Proof. We shall construct a sequence $\{A_k\}_{k=1}^{\infty}$ of finite sets such that A_k is a generalized Sidon set of order $h - 1$ for all $k \geq 1$, and $A = \cup_{k=1}^{\infty} A_k$ is an asymptotic basis of order h for $\mathbf{Z}$ whose representation function is equal to f.

Let $U = \{u_k\}_{k=1}^{\infty}$ be a sequence of integers such that

$$\mathrm{card}\left(\{k \in \mathbf{N} : u_k = n\}\right) = f(n)$$

for all integers n. It suffices to construct finite sets A_k such that for all integers n, we have

$$r_{A_k,h}(n) \leq f(n) \tag{4}$$

and

$$r_{A_k,h}(n) \geq \mathrm{card}\left(\{i \in \{1, 2, \ldots, k\} : u_i = n\}\right). \tag{5}$$

Choose positive integers d_1 and c_1 such that

$$f^{-1}(0) \subseteq [-d_1, d_1]$$

and

$$c_1 > 2h(d_1 + |u_1|).$$

By Lemma 1, the set

$$A_1 = D_{c_1, u_1} = \{-c_1, (h - 1)c_1 + u_1\}$$

is a generalized Sidon set of order h and $u_1 \in hA_1$. We shall prove that $hA_1 \cap f^{-1}(0) = \emptyset$. If $x \in f^{-1}(0)$, then $|x| \leq d_1$, and so $|x - u_1| \leq d_1 + |u_1|$. Again by Lemma 1, if $x \in hA_1 \setminus \{u_1\}$, then

$$|x - u_1| > \frac{c_1}{2} > h(d_1 + |u_1|) \geq 2(d_1 + |u_1|).$$

It follows that $hA_1 \cap f^{-1}(0) = \emptyset$, and so $r_{A_1,h}(n) \leq 1 \leq f(n)$ for all $n \in hA_1$ and $r_{A_1,h}(u_1) = 1$. Thus, the set A_1 satisfies conditions (4) and (5).

Let $k \geq 2$, and assume that we have constructed a generalized Sidon set A_{k-1} of order $h - 1$ that satisfies conditions (4) and (5). Choose positive integers d_k and c_k such that

$$f^{-1}(0) \cup \bigcup_{r=1}^{h} r A_{k-1} \subseteq [-d_k, d_k]$$

and

$$c_k > 2h(2d_k + |u_k|).$$

Let

$$A_k = A_{k-1} \cup D_{c_k, u_k} = A_{k-1} \cup \{-c_k, (h-1)c_k + u_k\}.$$

Then

$$hA_k = hA_{k-1} \cup \bigcup_{r=1}^{h} (r D_{c_k, u_k} + (h-r)A_{k-1}).$$

By Lemma 1, the set D_{c_k, u_k} is a generalized Sidon set of order h, and so every integer in the set $\bigcup_{r=1}^{h} r D_{c_k, u_k}$ has exactly one representation as the sum of at most h elements of D_{c_k, u_k}. Also, the minimum distance between the elements of $\bigcup_{r=1}^{h} r D_{c_k, u_k}$ is greater than $c_k/2$.

Let $x, x' \in \bigcup_{r=1}^{h} r D_{c_k, u_k}$ with $x \neq x'$. By Lemma 1, there are unique positive integers r, r' such that $x \in r D_{c_k, u_k}$ and $x' \in r' D_{c_k, u_k}$. If $y \in (h - r)A_{k-1}$ and $y' \in (h - r')A_{k-1}$, then

$$|y - y'| \leq |y| + |y'| \leq 2d_k < \frac{c_k}{2} \leq |x' - x|,$$

and so $x + y \neq x' + y'$. It follows that the sets $\{x\} + (h-r)A_{k-1}$ and $\{x'\} + (h-r')A_{k-1}$ are pairwise disjoint. Since A_{k-1} is a generalized Sidon set of order $h - 1$, it follows that every element of

$$\bigcup_{r=1}^{h} (r D_{c_k, u_k} + (h - r)A_{k-1})$$

has a unique representation as the sum of exactly h elements of A_k.

Recall that $u_k \in h D_{c_k, u_k}$ and $h A_{k-1} \cup f^{-1}(0) \subseteq [-d_k, d_k]$. If $w \in h A_{k-1} \cup f^{-1}(0)$, then $|u_k - w| \leq d_k + |u_k|$. If

$$z \in \bigcup_{r=1}^{h} (r D_{c_k, u_k} + (h - r)A_{k-1}),$$

then $z = x + y$, where $x \in r D_{c_k, u_k}$ for some $r \in [1, h]$ and $y \in (h - r)A_{k-1}$. If $z \neq u_k$, then $x \neq u_k$. It follows again from Lemma 1 that $|x - u_k| \geq c_k/2$ and

$$|z - w| = |x + y - w| = |x - u_k + u_k + y - w|$$

$$\geq |x - u_k| - |u_k + y - w|$$

$$\geq \frac{c_k}{2} - (2d_k + |u_k|)$$

$$> (h - 1)(2d_k + |u_k|)$$

$$> 0.$$

Therefore,

$$hA_k \subseteq \mathbf{Z} \setminus f^{-1}(0)$$

and

$$hA_{k-1} \cap \left(\bigcup_{r=1}^{h} (r D_{c_k,u_k} + (h - r)A_{k-1}) \right) = \emptyset \text{ or } \{u_k\}.$$

It follows that

$$r_{A_k,h}(n) = \begin{cases} r_{A_{k-1},h}(n) & \text{if } n \in hA_{k-1} \setminus \{u_k\}, \\ r_{A_{k-1},h}(u_k) + 1 & \text{if } n = u_k, \\ 1 & \text{if } n \in hA_k \setminus hA_{k-1}, \end{cases}$$

and so the set A_k satisfies conditions (4) and (5).

A similar argument shows that A_k is a generalized Sidon set of order $h - 1$. Let

$$Z = \bigcup_{h'=1}^{h-1} h' A_k = \bigcup_{h'=1}^{h-1} \left(\bigcup_{\substack{r,s=0 \\ r+s=h'}}^{h'} (r D_{c_k,u_k} + s A_{k-1}) \right)$$

$$= \bigcup_{\substack{r,s=0 \\ 1 \leq r+s \leq h-1}}^{h-1} (r D_{c_k,u_k} + s A_{k-1}).$$

Suppose that

$$z = x + y = x' + y' \in Z,$$

where $x \in r D_{c_k,u_k}, y \in s A_{k-1}, x' \in r' D_{c_k,u_k}, y' \in s' A_{k-1}$ for nonnegative integers r, s, r', s' such that $1 \leq r + s \leq r' + s' \leq h - 1$. If $x \neq x'$, then

$$|x - x'| \geq \frac{c_k}{2} > 2d_k \geq |y' - y|,$$

and so $x - x' \neq y' - y$, which is absurd. Therefore, $x = x'$ and $y = y'$. Since D_{c_k,u_k} is a generalized Sidon set of order h and A_{k-1} is a generalized Sidon set of order $h - 1$, it follows that x and y have *unique* representations as sums of at most $h - 1$ elements of D_{c_k,u_k} and A_{k-1}, respectively, and so z has a unique representation as the sum of at most $h - 1$ elements of A_k. This completes the proof. $\qquad \square$

By Theorem 7, for every function $f \in \mathcal{F}_0(\mathbf{Z})$, there exist infinitely many asymptotic bases A of order h such that $r_{A,h} = f$, and such bases can be constructed that are arbitrarily sparse. An open problem is to determine how dense such a set can be. Nathanson and Cilleruelo [2, 3] proved that for every $f \in \mathcal{F}_0$ and every $\varepsilon > 0$, there is a set A of integers with $r_{A,h} = f$ and

$$A(-x, x) \gg x^{\sqrt{2}-1-\varepsilon}$$

for all $x \geq 1$. The construction uses dense Sidon sets.

Problem 11. Let α_2^* be the supremum of the set of all positive real numbers α such that for every $f \in \mathcal{F}_0$, there is a set A of integers with $r_{A,h} = f$ and $A(-x, x) \gg x^\alpha$ for all $x \geq 1$. Determine α_2^*.

Problem 12. Let $h \geq 3$. Does there exist a positive real number α_h such that for every $f \in \mathcal{F}_0$, there is a set A of integers with $r_{A,h} = f$ and $A(-x, x) \gg x^{\alpha_h}$ for all $x \geq 1$. How large can α_h be?

We can extend the inverse problem for representation functions to functions $f : \mathbf{Z} \to \mathbf{N}_0 \cup \{\infty\}$ that have infinitely many zeros. In the case $h = 2$, if $f^{-1}(0)$ is a set of integers of density 0, then there we can construct a set A with $f = r_{A,2}$. The problem is open for higher orders h.

Problem 13. Let $f : \mathbf{Z} \to \mathbf{N}_0 \cup \{\infty\}$ be a function such that $d(f^{-1}(0)) = 0$. Let $h \geq 3$. Does there exist a set A of integers such that $r_{A,h}(n) = f(n)$ for all integers n?

We can extend this problem to functions whose zero sets have small positive density.

Problem 14. Let $h \geq 2$. Does there exist $\delta = \delta(h) > 0$ such that if $f : \mathbf{Z} \to \mathbf{N}_0 \cup \{\infty\}$ is a function with $d_U(f^{-1}(0)) < \delta$, then there exists a set A of integers such that $r_{A,h}(n) = f(n)$ for all integers n?

8 Representation Functions for Abelian Semigroups

The significant difference between inverse problems for $\mathbf{N}_0$ and $\mathbf{Z}$ derives in part from the fact that $\mathbf{Z}$ is a group but $\mathbf{N}_0$ is not. Nathanson [16] obtained some general inverse theorems for representation functions of "semigroups with a group component."

Let B be a subset of an abelian semigroup $\mathbf{X}$ and let $x \in \mathbf{X}$. We define the representation functions

$$r_{B,2}(x) = \mathrm{card}(\{\{b, b'\} \subseteq B : b + b' = x\})$$

and
$$\hat{r}_{B,2}(x) = \mathrm{card}(\{\{b, b'\} \subseteq B : b + b' = x \text{ and } b \neq b'\}).$$

We consider semigroups S with the property that $S + S = S$. Equivalently, for every $s \in S$ there exist $s', s'' \in S$ such that $s = s' + s''$. Every semigroup with identity has this property, since $s = s + 0$. There are also semigroups without identity that have this property. For example, if S is any totally ordered set without a smallest element, and if we define $s_1 + s_2 = \max(s_1, s_2)$, then S is an abelian semigroup such that $s = s + s$ for all $s \in S$, but S does not have an identity element.

Let S be an abelian semigroup and let $B \subseteq S$. For every positive integer h, we define the *dilation*

$$h * B = \{hb : b \in B\} = \{\underbrace{b + \cdots + b}_{h \text{ summands}} : b \in B\}.$$

Note that if G is an abelian group such that every element of G has order dividing h, then $h * G = \{0\}$.

Theorem 8. *Let S be a countable abelian semigroup such that for every $s \in S$ there exist $s', s'' \in S$ with $s = s' + s''$. Let G be a countably infinite abelian group such that the dilation $2 * G$ is infinite. Consider the abelian semigroup $\mathbf{X} = S \oplus G$ with projection map $\pi : \mathbf{X} \to G$. Let*

$$f : \mathbf{X} \to \mathbf{N}_0 \cup \{\infty\}$$

be any map such that the set $\pi(f^{-1}(0))$ is a finite subset of G. Then there exists a set $B \subseteq \mathbf{X}$ such that

$$\hat{r}_{B,2}(x) = f(x)$$

for all $x \in \mathbf{X}$.

Note that Theorem 8 is not true for all abelian semigroups. For example, let $\mathbf{N}$ be the additive semigroup of positive integers under addition, and $\mathbf{X} = \mathbf{N} \oplus \mathbf{Z}$. Since the equation $s' + s'' = 1$ has no solution in positive integers, it follows that for every set $B \subseteq \mathbf{X}$, we have $r_B(1, n) = \hat{r}_B(1, n) = 0$ for every $n \in \mathbf{Z}$. Thus, if $f : \mathbf{X} \to \mathbf{N}_0 \cup \{\infty\}$ is any function with $f(1, n) \neq 0$ for some integer n, then there does not exist a set $B \subseteq \mathbf{X}$ with $\hat{r}_{B,2} = f$.

Theorem 9. *Let G be a countably infinite abelian group such that the dilation $2 * G$ is infinite. Let*

$$f : G \to \mathbf{N}_0 \cup \{\infty\}$$

be any map such that $f^{-1}(0)$ is a finite subset of G. Then there exists a set B of order 2 for G such that

$$\hat{r}_{B,2}(x) = f(x)$$

for all $x \in \mathbf{X}$.

Theorem 10. *Let S be a countable abelian semigroup such that for every s ∈ S there exist s', s'' ∈ S with s = s' + s''. Let G be a countably infinite abelian group such that the dilation 12 * G is infinite. Consider the abelian semigroup* $\mathbf{X} = S \oplus G$ *with projection map* $\pi : \mathbf{X} \to G$. *Let*

$$f : \mathbf{X} \to \mathbf{N}_0 \cup \{\infty\}$$

be any map such that the set $\pi(f^{-1}(0))$ *is finite. Then there exists a set* $B \subseteq \mathbf{X}$ *such that*

$$r_{B,2}(x) = f(x)$$

for all $x \in \mathbf{X}$.

Theorem 11. *Let G be a countably infinite abelian group such that the dilation* $12 * G$ *is infinite. Let*

$$f : G \to \mathbf{N}_0 \cup \{\infty\}$$

be any map such that the set $f^{-1}(0)$ *is finite. Then there exists an asymptotic basis B of order 2 for G such that*

$$r_{B,2}(x) = f(x)$$

for all $x \in \mathbf{X}$.

The proofs of Theorems 8–11 can be found in [16].

Problem 15. What countable abelian semigroups $\mathbf{X}$ have the property that for every function $f : \mathbf{X} \to \mathbf{N}_0 \cup \{\infty\}$ such that the set $f^{-1}(0)$ is finite, there exists an asymptotic basis B of order 2 for $\mathbf{X}$ with $r_{B,2} = f$?

9 Bases Associated to Binary Linear Forms

Let $\Phi(x_1, x_2) = u_1 x_1 + u_2 x_2$ be a binary linear form with relatively prime integer coefficients u_1 and u_2. Let A_1 and A_2 be sets of integers. We define the set

$$\Phi(A_1, A_2) = \{\Phi(a_1, a_2) : a_1 \in A_1 \text{ and } a_2 \in A_2\}.$$

The *representation function* associated with the form Φ is

$$R_{A_1, A_2, \Phi}(n) = \mathrm{card}\left(\{(a_1, a_2) \in A_1 \times A_2 : \Phi(a_1, a_2) = n\}\right).$$

Then $R_{A_1, A_2, \Phi}$ is a function from $\mathbf{Z}$ into $\mathbf{N}_0 \cup \{\infty\}$. If $A_1 = A_2 = A$, we write

$$\Phi(A) = \Phi(A, A) = \{\Phi(a_1, a_2) : a_1, a_2 \in A\}$$

and

$$R_{A, \Phi}(n) = R_{A, A, \Phi}(n) = \mathrm{card}(\{(a_1, a_2) \in A^2 : \Phi(a_1, a_2) = n\}).$$

The set A will be a called a *unique representation basis with respect to the form* Φ if $R_{A, \Phi}(n) = 1$ for every integer n.

Lemma 2. *Let* $\Phi(x_1, x_2) = u_1 x_1 + u_2 x_2$ *be a binary linear form with relatively prime positive integer coefficients* $u_1 < u_2$. *Let A be a finite set of integers and let b be an integer. Then there exists a set C with $A \subseteq C$ and $|C \setminus A| = 2$ such that*

$$R_{C,\Phi}(n) = \begin{cases} R_{A,\Phi}(b) + 1 & \text{if } n = b, \\ R_{A,\Phi}(n) & \text{if } n \in \Phi(A) \setminus \{b\}, \\ 1 & \text{if } n \in \Phi(C) \setminus (\Phi(A) \cup \{b\}), \\ 0 & \text{if } n \notin \Phi(C). \end{cases} \tag{6}$$

Proof. Since $\gcd(u_1, u_2) = 1$, there exist integers v_1 and v_2 such that $\Phi(v_1, v_2) = u_1 v_1 + u_2 v_2 = 1$. Then

$$\Phi(bv_1 + u_2 t, bv_2 - u_1 t) = u_1(bv_1 + u_2 t) + u_2(bv_2 - u_1 t)$$
$$= b(u_1 v_1 + u_2 v_2) = b$$

for all integers t. Let $B = \{bv_1 + u_2 t, bv_2 - u_1 t\}$. If $t \neq (b(v_2 - v_1))/(u_1 + u_2)$, then $bv_1 + u_2 t \neq bv_2 - u_1 t$ and $|B| = 2$. We shall prove that there exist infinitely many integers t such that $A \cap B = \emptyset$ and the set $C = A \cup B$ satisfies conditions (6).

If $d = \max(\{|a| : a \in A\})$, then $|\Phi(a)| \leq (u_1 + u_2)d$ for all $a \in A$. The set $\Phi(C)$ is the union of the sets $\Phi(A)$, $\Phi(A, B)$, $\Phi(B, A)$, and $\Phi(B)$.

If $c \in \{\Phi(a, bv_1 + u_2 t) : a \in A\}$, then there exists $a \in A$ such that

$$c = u_1 a + u_2(bv_1 + u_2 t) = (u_1 a + u_2 v_1 b) + u_2^2 t,$$

and so $c > d$ for all sufficiently large integers t.

If $c' \in \{\Phi(a, bv_2 - u_1 t) : a \in A\}$, then there exists $a' \in A$ such that

$$c' = u_1 a' + u_2(bv_2 - u_1 t) = (u_1 a' + u_2 v_2 b) - u_1 u_2 t,$$

and so $c < -d$ for all sufficiently large integers t. Therefore,

$$\Phi(A) \cap \Phi(A, B) = \emptyset$$

for all sufficiently large integers t.

For every integer t, the functions $\Phi(a, bv_1 + u_2 t)$ and $\Phi(a, bv_2 - u_1 t)$ are strictly increasing functions of a. Moreover, there exist $a, a' \in A$ such that $\Phi(a, bv_1 + u_2 t) = \Phi(a', bv_2 - u_1 t)$ if and only if

$$(u_1 a + u_2 v_1 b) + u_2^2 t = (u_1 a' + u_2 v_2 b) - u_1 u_2 t,$$

that is, if and only if

$$(u_1 + u_2) u_2 t = u_1(a' - a) + u_2(v_2 - v_1)b,$$

and this identity holds only for finitely many t. Thus, for all sufficiently large t we have $R_{A,B,f}(n) \leq 1$ for all $n \in \mathbf{Z}$.

Similarly, if $c \in \{\Phi(bv_1 + u_2t, a) : a \in A\}$, then there exists $a \in A$ such that

$$c = u_1(bv_1 + u_2t) + u_2a = (u_1v_1b + u_2a) + u_1u_2t,$$

and so $c > d$ for all sufficiently large integers t.

If $c' \in \{\Phi(bv_2 - u_1t, a) : a \in A\}$, then there exists $a' \in A$ such that

$$c' = u_1(bv_2 - u_1t) + u_2a' = (u_1v_2b + u_2a') - u_1^2t,$$

and so $c < -d$ for all sufficiently large integers t. Therefore,

$$\Phi(A) \cap \Phi(B, A) = \emptyset$$

for all sufficiently large integers t. By the same method, we can prove that for all sufficiently large t we have $R_{B,A,f}(n) \leq 1$ for all $n \in \mathbf{Z}$ and

$$\Phi(A) \cap \Phi(B, A) = \Phi(A, B) \cap \Phi(B, A) = \emptyset.$$

Finally, the set $\Phi(B) = \{\Phi(b', b'') : b', b'' \in B\}$ consists of the integers b, $(u_1v_2 + u_2v_1)b + (u_2^2 - u_1^2)t$, $(u_1 + u_2)bv_1 + u_2(u_1 + u_2)t)$, and $(u_1 + u_2)bv_2 - u_1(u_1+u_2)t)$. The coefficients of t are the distinct integers $u_2^2 - u_1^2$, $u_2(u_1+u_2)$, and $-u_1(u_1 + u_2)$, and these are different from the numbers $-u_1u_2$, $-u_1^2$, u_1u_2, and u_2^2, which are the coefficients of t in $\Phi(A, B)$ and $\Phi(B, A)$. It follows that $|\Phi(B)| = 4$ and that the sets $\Phi(A)$, $\Phi(B, A)$, $\Phi(A, B)$, and $\Phi(B) \setminus \{b\}$ are pairwise disjoint for all sufficiently large t. This completes the proof.   □

Theorem 12. *Let* $\Phi(x_1, x_2) = u_1x_1 + u_2x_2$ *be a binary linear form with relatively prime positive integer coefficients* $u_1 < u_2$. *There exists a unique representation basis with respect to the form* Φ, *that is, a set* A *of integers such that* $R_{A,\Phi}(n) = 1$ *for all* $n \in \mathbf{Z}$.

Proof. We shall construct an increasing sequence of finite sets $A_1 \subseteq A_2 \subseteq \cdots$ such that $R_{A_k,f}(n) \leq 1$ for all $k \in \mathbf{N}$ and $n \in \mathbf{Z}$, and $A = \bigcup_{k=1}^{\infty} A_k$ is a unique representation basis for f. Let $A_1 = \{0, 1\}$. Then $\Phi(A_1) = \{0, u_1, u_2, u_1 + u_2\}$. Since $0 < u_1 < u_2 < u_1 + u_2$, it follows that $|\Phi(A_1)| = 4$ and $R_{A_1,f}(n) \leq 1$ for all $n \in \mathbf{Z}$.

Let A_k be a finite set of integers such that $R_{A_k,\Phi}(n) \leq 1$ for all $n \in \mathbf{Z}$. Let b be an integer such that

$$|b| = \min(\{|n| : n \notin \Phi(A_k)\}).$$

By Lemma 2, there is a set A_{k+1} containing A_k such that $b \in \Phi(A_{k+1})$ and $R_{A_{k+1},\Phi}(n) \leq 1$ for all $n \in \mathbf{Z}$. This completes the proof.   □

More general results about representation functions of binary linear forms appear in Nathanson [18].

Problem 16. Determine all m-ary linear forms $\Phi(x_1, \ldots, x_m) = u_1x_1 + \cdots + u_mx_m$ with nonzero, relatively prime integer coefficients such that there exists a unique representation basis with respect to Φ.

Problem 17. Let $m \geq 2$ and let $\Phi(x_1, \ldots, x_m)$ be an m-ary linear form with nonzero, relatively prime integer coefficients. Let $f : \mathbf{Z} \to \mathbf{N}_0 \cup \{\infty\}$ be a function such that $f^{-1}(0)$ is finite. Does there exist a set A such that $R_{A,\Phi} = f$?

Problem 18. Determine all m-ary linear forms Φ such that if A and B are sets of integers with $R_{A,\Phi} = R_{B,\Phi}$, then $A = B$.

Problem 19. Determine all m-ary linear forms Φ such that if A and B are finite sets of integers with $R_{A,\Phi} = R_{B,\Phi}$, then $A = B$.

The last problem is related to work of Ewell, Fraenkel, Gordon, Selfridge, and Straus [7, 8, 21].

References

1. Y.-G. Chen, *A problem on unique representation bases*, European J. Combin. **28** (2007), no. 1, 33–35.
2. J. Cilleruelo and M. B. Nathanson, *Dense sets of integers with prescribed representation functions*, preprint, 2007.
3. _____, *Perfect difference sets constructed from Sidon sets*, Combinatorica (to appear).
4. G. A. Dirac, *Note on a problem in additive number theory*, J. London Math. Soc. **26** (1951), 312–313.
5. P. Erdős and P. Turán, *On a problem of Sidon in additive number theory, and on some related problems*, J. London Math. Soc. **16** (1941), 212–215.
6. P. Erdős, *On the multiplicative representation of integers*, Israel J. Math. **2** (1964), 251–261.
7. J. A. Ewell, *On the determination of sets by sets of sums of fixed order*, Canad. J. Math. **20** (1968), 596–611.
8. B. Gordon, A. S. Fraenkel, and E. G. Straus, *On the determination of sets by the sets of sums of a certain order*, Pacific J. Math. **12** (1962), 187–196.
9. J. Lee, *Infinitely often dense bases with a prescribed representation function*, arXiv preprint, 2007.
10. J. C. M. Nash and M. B. Nathanson, *Cofinite subsets of asymptotic bases for the positive integers*, J. Number Theory **20** (1985), no. 3, 363–372.
11. M. B. Nathanson, *Representation functions of sequences in additive number theory*, Proc. Amer. Math. Soc. **72** (1978), no. 1, 16–20.
12. _____, *Multiplicative representations of integers*, Israel J. Math. **57** (1987), no. 2, 129–136.
13. _____, *Additive number theory: The classical bases*, Graduate Texts in Mathematics, vol. 164, Springer-Verlag, New York, 1996.
14. _____, *Unique representation bases for the integers*, Acta Arith. **108** (2003), no. 1, 1–8.
15. _____, *The inverse problem for representation functions of additive bases*, Number theory (New York, 2003), Springer, New York, 2004, pp. 253–262.
16. _____, *Representation functions of additive bases for abelian semigroups*, Int. J. Math. Math. Sci. (2004), no. 29-32, 1589–1597.
17. _____, *Every function is the representation function of an additive basis for the integers*, Port. Math. (N.S.) **62** (2005), no. 1, 55–72.

18. _____, *Representation functions of bases for binary linear forms*, Funct. Approx. Comment. Math. **37** (2007), 341–350.
19. J. Nešetřil and V. Rödl, *Two proofs in combinatorial number theory*, Proc. Amer. Math. Soc. **93** (1985), 185–188.
20. C. Sándor, *Partitions of natural numbers and their representation functions*, Integers **4** (2004), A18, 5 pp. (electronic).
21. J. L. Selfridge and E. G. Straus, *On the determination of numbers by their sums of a fixed order*, Pacific J. Math. **8** (1958), 847–856.
22. M. Tang and Y.-G. Chen, *A basis of* $\mathbb{Z}_m$, Colloq. Math. **104** (2006), no. 1, 99–103.

6

Mock Theta Functions, Ranks, and Maass Forms

Ken Ono

Department of Mathematics, University of Wisconsin, Madison, Wisconsin 53706
ono@math.wisc.edu

Summary. This paper is a survey of some recent joint work with Kathrin Bringmann on harmonic weak Maass forms. We summarize applications to Dyson's rank-partition statistic and the conjecture of Andrews and Dragonette on Ramanujan's third-order mock theta function $f(q)$.

Key words: Maass forms and mock theta functions.

2000 *Mathematics Subject Classifications*: 11P82, 05A17

1 Introduction

Generating functions play a central role throughout number theory. For example, in the theory of partitions, if $p(n)$ denotes the number of partitions of an integer n, then Euler observed that

$$P(q) := \sum_{n=0}^{\infty} p(n)q^{24n-1} = q^{-1} \prod_{n=1}^{\infty} \frac{1}{1-q^{24n}}$$

$$= q^{-1} + q^{23} + 2q^{47} + 3q^{71} + 5q^{95} + \cdots . \qquad (1.1)$$

Similarly, in the theory of quadratic forms, we have the following fundamental q-series identity of Jacobi:

$$\Theta(q) := \sum_{n=-\infty}^{\infty} q^{n^2} = \prod_{n=1}^{\infty} \frac{(1-q^{2n})^5}{(1-q^n)^2(1-q^{4n})^2} = 1 + 2q + 2q^4 + 2q^9 + \cdots . \quad (1.2)$$

Consequently, it follows that the integers $r_s(n)$, the number of representations of integers n as sums of s squares, are formally given as the coefficients of the q-series

The author thanks the National Science Foundation for its generous support, and he is grateful for the support of a Packard and a Romnes Fellowship.

K. Alladi (ed.), *Surveys in Number Theory*, DOI: 10.1007/978-0-387-78510-3_6,
© Springer Science+Business Media, LLC 2008

$$\sum_{n=0}^{\infty} r_s(n)q^n = \Theta(q)^s = \prod_{n=1}^{\infty} \frac{(1-q^{2n})^{5s}}{(1-q^n)^{2s}(1-q^{4n})^{2s}}.$$

As a third example, consider the q-series $A(q)$ defined by

$$A(q) := \sum_{n=1}^{\infty} a_E(n)q^n := q\prod_{n=1}^{\infty}(1-q^{4n})^2(1-q^{8n})^2$$

$$= q - 2q^5 - 3q^9 + 6q^{13} + 2q^{17} - \cdots. \tag{1.3}$$

It is well known that $L(E,s) = \sum_{n=1}^{\infty} \frac{a_E(n)}{n^s}$ is the Hasse–Weil L-function of the elliptic curve

$$E: \quad y^2 = x^3 - x.$$

In particular, this fact implies that if p is an odd prime, then

$$a_E(p) = p - \#\{(x,y) \in (\mathbb{F}_p)^2 : y^2 \equiv x^3 - x(\bmod\ p)\} = -\sum_{x=0}^{p-1}\left(\frac{x^3-x}{p}\right).$$

Therefore, the q-series $A(q)$ can be naively thought of as the "generating function" for the number of $\mathbb{F}_p$-points on the reductions of the elliptic curve E as one varies p.

These examples share the property that they all are generating functions that coincide with Fourier expansions of modular forms. Loosely speaking, a *modular form of weight k on a subgroup* $\Gamma \subset \mathrm{SL}_2(\mathbb{Z})$ *with multiplier system ϵ* is any meromorphic function $f(z)$ on $\mathbb{H}$ with the property that

$$f\left(\frac{az+b}{cz+d}\right) = \epsilon(a,b,c,d)(cz+d)^k f(z)$$

for every $\left(\begin{smallmatrix} a & b \\ c & d \end{smallmatrix}\right) \in \Gamma$. The modularity of these generating functions follows from the modularity of Dedekind's eta function

$$\eta(z) := q^{1/24}\prod_{n=1}^{\infty}(1-q^n) \tag{1.4}$$

(note: $q := e^{2\pi i z}$ throughout). More precisely, modularity follows from the well-known transformation laws

$$\eta(z+1) = e(1/24)\eta(z) \quad \text{and} \quad \eta(-1/z) = (-iz)^{\frac{1}{2}}\eta(z), \tag{1.5}$$

where $e(\alpha) := e^{2\pi i \alpha}$.

In this context, the theory of modular forms has played a central role in the study of partitions, quadratic forms, and elliptic curves, as well as many other topics throughout mathematics. The rich theory of modular forms allows one to prove theorems about asymptotics, congruences, and multiplicative relations.

On the other hand, there are many related questions in which modular forms do not appear to play a role. Here we consider problems on mock theta functions and partitions, and we show that *weak Maass forms* play a prominent role. We shall consider recent works [11, 12] by Bringmann and the author on mock theta functions and Dyson's partition ranks. In particular, we shall review recent results on:

- Modularity of mock theta functions,
- Dyson's partition ranks and partition congruences,
- The Andrews–Dragonette conjecture for the mock theta function $f(q)$.

2 Weak Maass Forms

We begin by recalling the notion of a weak Maass form of half-integral weight $k \in \frac{1}{2}\mathbb{Z} \setminus \mathbb{Z}$. If $z = x + iy$ with $x, y \in \mathbb{R}$, then the weight k hyperbolic Laplacian is given by

$$\Delta_k := -y^2 \left(\frac{\partial^2}{\partial x^2} + \frac{\partial^2}{\partial y^2} \right) + iky \left(\frac{\partial}{\partial x} + i \frac{\partial}{\partial y} \right). \tag{2.1}$$

If v is odd, then define ϵ_v by

$$\epsilon_v := \begin{cases} 1 & \text{if } v \equiv 1 \, (\text{mod } 4), \\ i & \text{if } v \equiv 3 \, (\text{mod } 4). \end{cases} \tag{2.2}$$

A *weak Maass form of weight k on a subgroup* $\Gamma \subset \Gamma_0(4)$ is any smooth function $f : \mathbb{H} \to \mathbb{C}$ satisfying the following:

(1) For all $A = \left(\begin{smallmatrix} a & b \\ c & d \end{smallmatrix} \right) \in \Gamma$ and all $z \in \mathbb{H}$, we have

$$f(Az) = \left(\frac{c}{d} \right)^{2k} \epsilon_d^{-2k} (cz + d)^k \, f(z).$$

Here $\left(\frac{c}{d} \right)$ denotes the extended Legendre symbol.
(2) We have that $\Delta_k f = 0$.
(3) The function $f(z)$ has at most linear exponential growth at all the cusps of Γ.

Similarly, we have the notion of a weak Maass form with Nebentypus. To define it, suppose that N is a positive integer, and that $\psi \, (\text{mod } 4N)$ is a Dirichlet character. A weight k weak Maass form on $\Gamma_1(4N)$ is a *weak Maass form on $\Gamma_0(4N)$ with Nebentypus character ψ* if for every $A = \left(\begin{smallmatrix} a & b \\ c & d \end{smallmatrix} \right) \in \Gamma_0(4N)$ and all $z \in \mathbb{H}$ we have

$$f(Az) = \psi(d) \left(\frac{c}{d} \right)^{2k} \epsilon_d^{-2k} (cz + d)^k \, f(z).$$

Remark. The transformation laws in these definitions coincide with those in Shimura's theory of half-integral-weight modular forms [31].

Weak Maass forms can be related to classical *weakly holomorphic modular forms*, those forms whose poles (if there are any) are supported at cusps. A weak Maass form that is holomorphic on $\mathbb{H}$ is already a weakly holomorphic modular form. More generally, one may relate weak Maass forms to weakly holomorphic modular forms using the antilinear differential operator ξ_k defined by

$$\xi_k(f)(z) := 2iy^k \overline{\frac{\partial}{\partial \bar{z}} f(z)}. \tag{2.3}$$

Here we have that

$$\frac{\partial}{\partial \bar{z}} = \frac{1}{2}\left(\frac{\partial}{\partial x} + i\frac{\partial}{\partial y}\right).$$

In their work on geometric theta lifts, Bruinier and Funke (see Proposition 3.2 of [13]) proved the following valuable proposition.

Proposition 2.1. *If $f(z)$ is a weak Maass form of weight k for the group $\Gamma_0(4N)$ with Nebentypus χ, then $\xi_k(f)$ is a weakly holomorphic modular form of weight $2 - k$ on $\Gamma_0(4N)$ with Nebentypus $\bar{\chi}$. Furthermore, ξ_k has the property that its kernel consists of those weight-k weak Maass forms that are weakly holomorphic modular forms.*

Remark. Proposition 2.1 holds for weak Maass forms on any subgroup $\Gamma \subset \Gamma_0(4)$, not just those with Nebentypus.

The purpose of this expository paper is to describe a number of instances in which the theory of weak Maass forms gives new results on mock theta functions and partition ranks. In particular, it will turn out that well-known generating functions appear as "pieces" of weak Maass forms. Weak Maass forms have Fourier expansions of the form

$$f(z) = \sum_{n=n_0}^{\infty} \gamma(f, n; y)q^{-n} + \sum_{n=n_1}^{\infty} a(f, n)q^n.$$

As one sees, the Fourier coefficients $\gamma(f, n; y)$ are functions in y, the imaginary part of z, while the coefficients $a(f, n)$ are ordinary complex numbers. Therefore, we shall refer to $\sum_{n=n_0}^{\infty} \gamma(f, n; y)q^{-n}$ as the "nonholomorphic part" of $f(z)$, and we shall refer to $\sum_{n=n_1}^{\infty} a(f, n)q^n$ as its "holomorphic part." The number-theoretic generating functions we consider are holomorphic parts of weak Maass forms.

3 Mock Theta Functions and Partition Ranks

The mock theta-functions give us tantalizing hints of a grand synthesis still to be discovered. Somehow it should be possible to build them into a coherent group-theoretical structure, analogous to the structure of modular forms which Hecke built around the old theta-functions of Jacobi. This remains a challenge for the future. My dream is that I will live to see the day when our young physicists, struggling to bring the predictions of superstring theory into correspondence with the facts of nature, will be led to enlarge their analytic machinery to include not only theta-functions but mock theta-functions. ... But before this can happen, the purely mathematical exploration of the mock-modular forms and their mock-symmetries must be carried a great deal further.

Freeman Dyson, 1987
Ramanujan Centenary Conference

Dyson's words (see page 20 of [18]) refers to 22 peculiar q-series, such as

$$f(q) := 1 + \sum_{n=1}^{\infty} \frac{q^{n^2}}{(1+q)^2(1+q^2)^2 \cdots (1+q^n)^2}, \tag{3.1}$$

which were defined by Ramanujan and Watson decades ago. In his last letter to Hardy, dated January 1920 (see pages 127–131 of [28]), Ramanujan lists 17 such functions, and he gives 2 more in his "Lost Notebook" [28]. In his paper "The final problem: an account of the mock theta functions" [33], Watson defines 3 further functions. Although much remains unknown about these enigmatic series, Ramanujan's mock theta functions have been the subject of an astonishing number of important works (for example, see [3, 4, 6, 7, 14, 15, 16, 20, 21, 22, 23, 28, 29, 33, 34, 35, 36] to name a few).

In his 2002 Ph.D. thesis [36], written under the direction of Zagier, Zwegers made an important step in the direction of Dyson's "challenge for the future" by relating many of Ramanujan's mock theta functions to real analytic vector-valued modular forms. More recently, Bringmann and the author have shown [12] that Dyson's own rank-generating function can already be used to construct the desired "coherent group-theoretical structure, analogous to the structure of modular forms which Hecke built around old theta functions of Jacobi." More precisely, we relate specializations of his partition rank-generating function to weak Maass forms, and we show that the nonholomorphic parts of these forms are period integrals of theta functions, thereby realizing Dyson's speculation that such a picture should involve theta functions.

In an effort to provide a combinatorial explanation of Ramanujan's congruences for $p(n)$, Dyson introduced [17] the so-called rank of a partition. The rank of a partition is defined to be its largest part minus the number of its parts. If $N(m, n)$ denotes the number of partitions of n with rank m, then it is well known that

$$R(w; q) := 1 + \sum_{n=1}^{\infty} \sum_{m=-\infty}^{\infty} N(m, n)w^m q^n = 1 + \sum_{n=1}^{\infty} \frac{q^{n^2}}{(wq; q)_n (w^{-1}q; q)_n}, \tag{3.2}$$

where

$$(a; q)_n := (1 - a)(1 - aq) \cdots (1 - aq^{n-1}),$$

$$(a; q)_\infty := \prod_{m=0}^{\infty} (1 - aq^m).$$

Letting $w = -1$, we obtain the series

$$R(-1; q) = 1 + \sum_{n=1}^{\infty} \frac{q^{n^2}}{(1+q)^2 (1+q^2)^2 \cdots (1+q^n)^2}.$$

This is the mock theta function $f(q)$ given in (3.1). This observation connects the additive number theory of partitions to mock theta functions. In the next section, we link the specializations of these generating functions, at roots of unity $w \neq 1$, to weight 1/2 weak Maass forms.

3.1 Dyson's Generating Functions and Maass Forms

Here we complete Dyson's generating functions to obtain weak Maass forms. Suppose that $0 < a < c$ are integers, and let $\zeta_c := e^{2\pi i/c}$. If $f_c := \frac{2c}{\gcd(c,6)}$, then define the theta function $\Theta\left(\frac{a}{c}; \tau\right)$ by

$$\Theta\left(\frac{a}{c}; \tau\right) := \sum_{m \,(\mathrm{mod}\, f_c)} (-1)^m \sin\left(\frac{a\pi(6m+1)}{c}\right) \cdot \theta\left(6m+1, 6f_c; \frac{\tau}{24}\right), \quad (3.3)$$

where $\tau \in \mathbb{H}$ and

$$\theta(\alpha, \beta; \tau) := \sum_{n \equiv \alpha \,(\mathrm{mod}\, \beta)} n e^{2\pi i \tau n^2}. \quad (3.4)$$

Throughout, let $\ell_c := \mathrm{lcm}(2c^2, 24)$, and let $\tilde{\ell}_c := \ell_c/24$. It is well known [31] that $\Theta\left(\frac{a}{c}; \ell_c \tau\right)$ is a weight 3/2 cusp form, and we use it to define the function $S_1\left(\frac{a}{c}; z\right)$:

$$S_1\left(\frac{a}{c}; z\right) := \frac{-i \sin\left(\frac{\pi a}{c}\right) \ell_c^{\frac{1}{2}}}{\sqrt{3}} \int_{-\bar{z}}^{i\infty} \frac{\Theta\left(\frac{a}{c}; \ell_c \tau\right)}{\sqrt{-i(\tau + z)}} \, d\tau. \quad (3.5)$$

Using this notation, define $D\left(\frac{a}{c}; z\right)$ by

$$D\left(\frac{a}{c}; z\right) := -S_1\left(\frac{a}{c}; z\right) + q^{-\frac{\ell_c}{24}} R(\zeta_c^a; q^{\ell_c}). \quad (3.6)$$

Theorem 3.1 (Bringmann–Ono, Theorems 1.1 and 1.2 of [12]). *If $0 < a < c$, then $D\left(\frac{a}{c}; z\right)$ is a weight 1/2 weak Maass form on Γ_c, where*

$$\Gamma_c := \left\langle \begin{pmatrix} 1 & 1 \\ 0 & 1 \end{pmatrix}, \begin{pmatrix} 1 & 0 \\ \ell_c^2 & 1 \end{pmatrix} \right\rangle. \quad (3.7)$$

Moreover, if c is odd, then $D\left(\frac{a}{c}; z\right)$ is a weak Maass form of weight 1/2 on $\Gamma_1(144 f_c^2 \tilde{\ell}_c)$.

Sketch of the proof. The conclusion of the theorem follows from a general result (see Theorem 3.4 of [12]) about vector-valued weight 1/2 weak Maass forms for the modular group $\mathrm{SL}_2(\mathbb{Z})$, a result that is of independent interest. For brevity, here we only sketch the proof of the first claim.

To prove this claim, we require modular transformation laws for the series $R(\zeta_c^a; q^{\ell_c})$. Some of these laws have been obtained recently by Gordon and McIntosh [21]. To state their results, we first define the series

$$M\left(\frac{a}{c}; z\right) = M\left(\frac{a}{c}; q\right) := \frac{1}{(q; q)_\infty} \sum_{n=-\infty}^{\infty} \frac{(-1)^n q^{n+\frac{a}{c}}}{1 - q^{n+\frac{a}{c}}} \cdot q^{\frac{3}{2}n(n+1)},$$

$$M_1\left(\frac{a}{c}; z\right) = M_1\left(\frac{a}{c}; q\right) := \frac{1}{(q; q)_\infty} \sum_{n=-\infty}^{\infty} \frac{(-1)^{n+1} q^{n+\frac{a}{c}}}{1 + q^{n+\frac{a}{c}}} \cdot q^{\frac{3}{2}n(n+1)},$$

$$N\left(\frac{a}{c}; z\right) = N\left(\frac{a}{c}; q\right) := \frac{1}{(q; q)_\infty}$$
$$\times \left(1 + \sum_{n=1}^{\infty} \frac{(-1)^n (1 + q^n)\left(2 - 2\cos\left(\frac{2\pi a}{c}\right)\right)}{1 - 2q^n \cos\left(\frac{2\pi a}{c}\right) + q^{2n}} \cdot q^{\frac{n(3n+1)}{2}}\right),$$

$$N_1\left(\frac{a}{c}; z\right) = N_1\left(\frac{a}{c}; q\right) := \frac{1}{(q; q)_\infty}$$
$$\times \sum_{n=0}^{\infty} \frac{(-1)^n (1 - q^{2n+1})}{1 - 2q^{n+\frac{1}{2}} \cos\left(\frac{2\pi a}{c}\right) + q^{2n+1}} \cdot q^{\frac{3n(n+1)}{2}}. \tag{3.8}$$

Gordon and McIntosh prove the following q-series identities:

$$M\left(\frac{a}{c}; q\right) = \sum_{n=1}^{\infty} \frac{q^{n(n-1)}}{(q^{\frac{a}{c}}; q)_n \cdot (q^{1-\frac{a}{c}}; q)_n}, \tag{3.9}$$

$$N\left(\frac{a}{c}; q\right) = 1 + \sum_{n=1}^{\infty} \frac{q^{n^2}}{\prod_{j=1}^{n}\left(1 - 2\cos\left(\frac{2\pi a}{c}\right)q^j + q^{2j}\right)}. \tag{3.10}$$

Obviously, (3.8) and (3.10) imply the important fact that

$$R(\zeta_c^a; q) = N\left(\frac{a}{c}; q\right). \tag{3.11}$$

Their transformation laws involve the following Mordell integrals:

$$J\left(\frac{a}{c}; \alpha\right) := \int_0^\infty e^{-\frac{3}{2}\alpha x^2} \cdot \frac{\cosh\left(\left(\frac{3a}{c} - 2\right)\alpha x\right) + \cosh\left(\left(\frac{3a}{c} - 1\right)\alpha x\right)}{\cosh(3\alpha x/2)} \, dx,$$

$$J_1\left(\frac{a}{c}; \alpha\right) := \int_0^\infty e^{-\frac{3}{2}\alpha x^2} \cdot \frac{\sinh\left(\left(\frac{3a}{c} - 2\right)\alpha x\right) - \sinh\left(\left(\frac{3a}{c} - 1\right)\alpha x\right)}{\sinh(3\alpha x/2)} \, dx. \tag{3.12}$$

Suppose that α and β have the property that $\alpha\beta = \pi^2$. Gordon and McIntosh then prove (see page 199 of [21]) that if $q := e^{-\alpha}$ and $q_1 := e^{-\beta}$, then

$$q^{\frac{3a}{2c}(1-\frac{a}{c})-\frac{1}{24}} \cdot M\left(\frac{a}{c}; q\right) = \sqrt{\frac{\pi}{2\alpha}} \csc\left(\frac{a\pi}{c}\right) q_1^{-\frac{1}{6}} \cdot N\left(\frac{a}{c}; q_1^4\right)$$

$$- \sqrt{\frac{3\alpha}{2\pi}} \cdot J\left(\frac{a}{c}; \alpha\right),$$

$$q^{\frac{3a}{2c}(1-\frac{a}{c})-\frac{1}{24}} \cdot M_1\left(\frac{a}{c}; q\right) = -\sqrt{\frac{2\pi}{\alpha}} q_1^{\frac{4}{3}} \cdot N_1\left(\frac{a}{c}; q_1^2\right) - \sqrt{\frac{3\alpha}{2\pi}} \cdot J_1\left(\frac{a}{c}; \alpha\right).$$

$$(3.13)$$

Using the functions

$$\mathcal{N}\left(\frac{a}{c}; q\right) = \mathcal{N}\left(\frac{a}{c}; z\right) := \csc\left(\frac{a\pi}{c}\right) \cdot q^{-\frac{1}{24}} \cdot N\left(\frac{a}{c}; q\right), \qquad (3.14)$$

$$\mathcal{M}\left(\frac{a}{c}; q\right) = \mathcal{M}\left(\frac{a}{c}; z\right) := 2q^{\frac{3a}{2c}\cdot(1-\frac{a}{c})-\frac{1}{24}} \cdot M\left(\frac{a}{c}; q\right), \qquad (3.15)$$

define the vector-valued function $F\left(\frac{a}{c}; z\right)$ by

$$F\left(\frac{a}{c}; z\right) := \left(F_1\left(\frac{a}{c}; z\right), F_2\left(\frac{a}{c}; z\right)\right)^T$$

$$= \left(\sin\left(\frac{\pi a}{c}\right) \mathcal{N}\left(\frac{a}{c}; \ell_c z\right), \sin\left(\frac{\pi a}{c}\right) \mathcal{M}\left(\frac{a}{c}; \ell_c z\right)\right)^T. \qquad (3.16)$$

Similarly, define the vector-valued (nonholomorphic) function $G\left(\frac{a}{c}; z\right)$ by

$$G\left(\frac{a}{c}; z\right) = \left(G_1\left(\frac{a}{c}; z\right), G_2\left(\frac{a}{c}; z\right)\right)^T$$

$$:= \left(2\sqrt{3}\sin\left(\frac{\pi a}{c}\right) \sqrt{-i\ell_c z} \cdot J\left(\frac{a}{c}; -2\pi i\ell_c z\right),\right.$$

$$\left.\frac{2\sqrt{3}\sin\left(\frac{\pi a}{c}\right)}{i\ell_c z} \cdot J\left(\frac{a}{c}; \frac{2\pi i}{\ell_c z}\right)\right)^T. \qquad (3.17)$$

The transformations in (3.13) easily imply the following transformations under the generators of Γ_c.

Lemma 3.2. *Assume the notation and hypotheses above. For $z \in \mathbb{H}$, we have*

$$F\left(\frac{a}{c}; z+1\right) = F\left(\frac{a}{c}; z\right),$$

$$\frac{1}{\sqrt{-i\ell_c z}} \cdot F\left(\frac{a}{c}; -\frac{1}{\ell_c^2 z}\right) = \begin{pmatrix} 0 & 1 \\ 1 & 0 \end{pmatrix} \cdot F\left(\frac{a}{c}; z\right) + G\left(\frac{a}{c}; z\right).$$

The Mordell vector $G\left(\frac{a}{c}; z\right)$ arises as integrals of the theta function $\Theta\left(\frac{a}{c}; \tau\right)$.

Lemma 3.3. *Assume the notation and hypotheses above. For $z \in \mathbb{H}$, we have*

$$G\left(\frac{a}{c}; z\right) = \frac{i\ell_c^{\frac{1}{2}} \sin\left(\frac{\pi a}{c}\right)}{\sqrt{3}} \int_0^{i\infty} \frac{\left((-i\ell_c\tau)^{-\frac{3}{2}} \Theta\left(\frac{a}{c}; -\frac{1}{\ell_c\tau}\right), \Theta\left(\frac{a}{c}; \ell_c\tau\right), \right)^T}{\sqrt{-i(\tau+z)}} d\tau.$$

Proof of the lemma. For brevity, we prove the asserted formula only for the first component of $G\left(\frac{a}{c}; z\right)$. The proof for the second component follows in the same way.

By analytic continuation and a change of variables (note: we may assume that $z = it$ with $t > 0$), we find that

$$J\left(\frac{a}{c}; \frac{2\pi}{\ell_c t}\right)$$

$$= \ell_c t \cdot \int_0^\infty e^{-3\ell_c\pi t x^2} \cdot \frac{\cosh\left(\left(\frac{3a}{c} - 2\right) 2\pi x\right) + \cosh\left(\left(\frac{3a}{c} - 1\right) 2\pi x\right)}{\cosh(3\pi x)} dx.$$

Using the Mittag-Leffler theory of partial fraction decompositions, one finds that

$$\frac{\cosh\left(\left(\frac{3a}{c} - 2\right) 2\pi x\right) + \cosh\left(\left(\frac{3a}{c} - 1\right) 2\pi x\right)}{\cosh(3\pi x)}$$

$$= \frac{-i}{\sqrt{3}\pi} \sum_{n\in\mathbb{Z}} \frac{(-1)^n \sin\left(\frac{\pi a(6n+1)}{c}\right)}{x - i\left(n + \frac{1}{6}\right)} - \frac{i}{\sqrt{3}\pi} \sum_{n\in\mathbb{Z}} \frac{(-1)^n \sin\left(\frac{\pi a(6n+1)}{c}\right)}{-x - i\left(n + \frac{1}{6}\right)}.$$

By introducing the extra term $\frac{1}{i\left(n+\frac{1}{6}\right)}$, we just have to consider

$$\int_{-\infty}^\infty e^{-3\pi\ell_c t x^2} \sum_{n\in\mathbb{Z}} (-1)^n \sin\left(\frac{\pi a(6n + 1)}{c}\right) \left(\frac{1}{x - i\left(n + \frac{1}{6}\right)} + \frac{1}{i\left(n + \frac{1}{6}\right)}\right) dx.$$

Since this expression is absolutely convergent, we may interchange summation and integration to obtain

$$J\left(\frac{a}{c}; \frac{2\pi}{\ell_c t}\right) = \frac{-\ell_c it}{\sqrt{3}\pi} \sum_{n\in\mathbb{Z}} (-1)^n \sin\left(\frac{\pi a(6n + 1)}{c}\right) \int_{-\infty}^\infty \frac{e^{-3\pi\ell_c t x^2}}{x - i\left(n + \frac{1}{6}\right)} dx.$$

For all $s \in \mathbb{R} \setminus \{0\}$, we have the identity

$$\int_{-\infty}^\infty \frac{e^{-\pi t x^2}}{x - is} dx = \pi i s \int_0^\infty \frac{e^{-\pi u s^2}}{\sqrt{u+t}} du$$

(this follows since both sides are solutions of $\left(-\frac{\partial}{\partial t} + \pi s^2\right) f(t) = \frac{\pi i s}{\sqrt{t}}$ and have the same limit 0 as $t \mapsto \infty$ and hence are equal). Hence we may conclude that

$$J\left(\frac{a}{c}; \frac{2\pi}{\ell_c t}\right)$$

$$= \frac{\ell_c t}{6\sqrt{3}} \sum_{n \in \mathbb{Z}} (-1)^n (6n+1) \sin\left(\frac{\pi a(6n+1)}{c}\right) \int_0^\infty \frac{e^{-\pi(n+1/6)^2 u}}{\sqrt{u + 3\ell_c t}} \, du.$$

Substituting $u = -3\ell_c i \tau$, and interchanging summation and integration gives

$$J\left(\frac{a}{c}; \frac{2\pi}{\ell_c t}\right)$$

$$= \frac{-it\ell_c^{\frac{3}{2}}}{6} \int_0^{i\infty} \frac{\sum_{n \in \mathbb{Z}} (-1)^n (6n+1) \sin\left(\frac{\pi a(6n+1)}{c}\right) e^{3\pi i \ell_c \tau \left(n + \frac{1}{6}\right)^2}}{\sqrt{-i(\tau + it)}} \, d\tau.$$

The claim follows since the sum over n coincides with definition (3.3). □

We must determine the necessary modular transformation properties of the vector

$$S\left(\frac{a}{c}; z\right) = \left(S_1\left(\frac{a}{c}; z\right), S_2\left(\frac{a}{c}; z\right)\right)$$

$$:= \frac{-i \sin\left(\frac{\pi a}{c}\right) \ell_c^{\frac{1}{2}}}{\sqrt{3}} \int_{-\bar{z}}^{i\infty} \frac{\left(\Theta\left(\frac{a}{c}; \ell_c \tau\right), (-i\ell_c \tau)^{-\frac{3}{2}} \Theta\left(\frac{a}{c}; -\frac{1}{\ell_c \tau}\right)\right)^T}{\sqrt{-i(\tau + z)}} \, d\tau.$$
$$(3.18)$$

Since $\Theta\left(\frac{a}{c}; \ell_c z\right)$ is a cusp form, the integral above is absolutely convergent. The next lemma shows that $S\left(\frac{a}{c}; z\right)$ satisfies the same transformations as $F\left(\frac{a}{c}; z\right)$.

Lemma 3.4. *Assume the notation and hypotheses above. For $z \in \mathbb{H}$, we have*

$$S\left(\frac{a}{c}; z+1\right) = S\left(\frac{a}{c}; z\right),$$

$$\frac{1}{\sqrt{-i\ell_c z}} \cdot S\left(\frac{a}{c}; -\frac{1}{\ell_c^2 z}\right) = \begin{pmatrix} 0 & 1 \\ 1 & 0 \end{pmatrix} \cdot S\left(\frac{a}{c}; z\right) + G\left(\frac{a}{c}; z\right).$$

Proof of the lemma. Using the Fourier expansion of $\Theta\left(\frac{a}{c}; z\right)$, one easily sees that

$$S_1\left(\frac{a}{c}; z+1\right) = S_1\left(\frac{a}{c}; z\right).$$

Using classical facts about theta functions [31], we also have that

$$S_2\left(\frac{a}{c}; z+1\right) = S_2\left(\frac{a}{c}; z\right).$$

Hence, it suffices to prove the second transformation law. We directly compute

$$\frac{1}{\sqrt{-i\ell_c z}} \cdot S\left(\frac{a}{c}; -\frac{1}{\ell_c^2 z}\right)$$

$$= \frac{i \sin\left(\frac{\pi a}{c}\right) \ell_c^{\frac{1}{2}}}{\sqrt{3}\sqrt{-i\ell_c z}} \int_{\frac{1}{\ell_c^2 \bar{z}}}^{i\infty} \frac{\left(\Theta\left(\frac{a}{c}; \ell_c \tau\right), (-i\ell_c \tau)^{-\frac{3}{2}}\Theta\left(\frac{a}{c}; -\frac{1}{\ell_c \tau}\right)\right)^T}{\sqrt{-i\left(\tau - \frac{1}{\ell_c^2 z}\right)}} \, d\tau,$$

and after making the change of variable $\tau \mapsto -\frac{1}{\ell_c^2 \tau}$, we obtain

$$\frac{1}{\sqrt{-i\ell_c z}} \cdot S\left(\frac{a}{c}; -\frac{1}{\ell_c^2 z}\right)$$

$$= \frac{i \sin\left(\frac{\pi a}{c}\right) \ell_c^{\frac{1}{2}}}{\sqrt{3}} \int_0^{-\bar{z}} \frac{\left((-i\ell_c \tau)^{-\frac{3}{2}}\Theta\left(\frac{a}{c}; -\frac{1}{\ell_c \tau}\right), \Theta\left(\frac{a}{c}, \ell_c \tau\right)\right)^T}{\sqrt{-i(\tau + z)}} \, d\tau.$$

Consequently, we obtain the desired conclusion

$$\frac{1}{\sqrt{-i\ell_c z}} \cdot S\left(\frac{a}{c}; -\frac{1}{\ell_c^2 z}\right) - \begin{pmatrix} 0 & 1 \\ 1 & 0 \end{pmatrix} \cdot S\left(\frac{a}{c}; z\right)$$

$$= \frac{i \sin\left(\frac{\pi a}{c}\right) \ell_c^{\frac{1}{2}}}{\sqrt{3}} \int_0^{i\infty} \frac{\left((-i\ell_c \tau)^{-\frac{3}{2}}\Theta\left(\frac{a}{c}; -\frac{1}{\ell_c \tau}\right), \Theta\left(\frac{a}{c}; \ell_c \tau\right)\right)^T}{\sqrt{-i(\tau + z)}} \, d\tau$$

$$= G\left(\frac{a}{c}; z\right).$$

$\square$

Using (3.6), (3.8), (3.11), (3.14), and (3.16), we find that we have already determined the transformation laws satisfied by $D\left(\frac{a}{c}; z\right)$, since we have

$$\begin{pmatrix} 1 & 0 \\ \ell_c^2 & 1 \end{pmatrix} = \begin{pmatrix} 0 & 1 \\ -\ell_c^2 & 0 \end{pmatrix} \begin{pmatrix} 1 & -1 \\ 0 & 1 \end{pmatrix} \begin{pmatrix} 0 & -\frac{1}{\ell_c^2} \\ 1 & 0 \end{pmatrix}.$$

The key point is that the first and third matrices on the right provide the same Möbius transformation on $\mathbb{H}$. Therefore the transformation laws for $D\left(\frac{a}{c}; z\right)$ follow from Lemma 3.2 and Lemma 3.4.

Now we show that $D\left(\frac{a}{c}; z\right)$ is annihilated by

$$\Delta_{\frac{1}{2}} = -y^2 \left(\frac{\partial^2}{\partial x^2} + \frac{\partial^2}{\partial y^2}\right) + \frac{iy}{2}\left(\frac{\partial}{\partial x} + i\frac{\partial}{\partial y}\right) = -4y^{\frac{3}{2}}\frac{\partial}{\partial z}\sqrt{y}\frac{\partial}{\partial \bar{z}}.$$

Since $q^{-\frac{\ell_c}{24}} R(\zeta_b^a; q^{\ell_c})$ is a holomorphic function in z, we get

$$\frac{\partial}{\partial \bar{z}}\left(D\left(\frac{a}{c}; z\right)\right) = -\frac{\partial}{\partial \bar{z}}\left(S_1\left(\frac{a}{c}; z\right)\right) = \frac{\sin\left(\frac{\pi a}{c}\right)\ell_c^{\frac{1}{2}}}{\sqrt{6y}} \cdot \Theta\left(\frac{a}{c}; -\ell_c \bar{z}\right).$$

Hence, we find that $\sqrt{y}\frac{\partial}{\partial \bar{z}}\left(D\left(\frac{a}{c}; z\right)\right)$ is antiholomorphic, and so

$$\frac{\partial}{\partial z}\sqrt{y}\frac{\partial}{\partial \bar{z}}\left(D\left(\frac{a}{c}; z\right)\right) = 0.$$

To complete the proof, it suffices to show that $D\left(\frac{a}{c}; z\right)$ has at most linear exponential growth at cusps. This follows from the convergence of the period integral $S_1\left(\frac{a}{c}; z\right)$ and the transformation laws satisfied by $D\left(\frac{a}{c}; z\right)$.   □

Remark. Zwegers observed [35] that one may interpret Watson's transformation laws for the third-order mock theta functions in terms of a vector-valued real analytic modular form on $SL_2(\mathbb{Z})$. Bringmann and the author constructed (see Theorem 3.4 of [12]) an infinite class of $SL_2(\mathbb{Z})$ vector-valued weight-1/2 weak Maass forms. This result implies Theorem 3.1, and it is a generalization of Zwegers' observation.

3.2 Ramanujan's Partition Congruences

Theorem 3.1 sheds new light on the role that Dyson's rank plays in the theory of partition congruences. If r and t are integers, then let $N(r, t; n)$ be the number of partitions of n whose rank is $r \pmod{t}$. Using a standard argument involving the orthogonality relations on sums of roots of unity, it is straightforward to deduce that if $0 \leq r < t$ are integers, then

$$\sum_{n=0}^{\infty} N(r, t; n)q^n = \frac{1}{t}\sum_{n=0}^{\infty} p(n)q^n + \frac{1}{t}\sum_{j=1}^{t-1}\zeta_t^{-rj} \cdot R(\zeta_t^j; q). \qquad (3.19)$$

By Theorem 3.1, it then follows that

$$\sum_{n=0}^{\infty}\left(N(r, t; n) - \frac{p(n)}{t}\right)q^{\ell_t n - \frac{\ell_t}{24}}$$

is the holomorphic part of a weak Maass form of weight 1/2 on Γ_t, one that is given as an appropriate weighted sum of the weak Maass forms $D\left(\frac{a}{t}; z\right)$. If t is odd, then it is on $\Gamma_1(144 f_t^2 \widetilde{\ell}_t)$. This result allows us to relate many "sieved" generating functions to weakly holomorphic modular forms.

Theorem 3.5 (Bringmann–Ono, Theorem 1.4 of [12]). *If $0 \leq r < t$ are integers, where t is odd, and $\mathcal{P} \nmid 6t$ is prime, then*

$$\sum_{\substack{n \geq 1 \\ \left(\frac{24\ell_t n - \ell_t}{\mathcal{P}}\right) = -\left(\frac{-24\widetilde{\ell}_t}{\mathcal{P}}\right)}}\left(N(r, t; n) - \frac{p(n)}{t}\right)q^{\ell_t n - \frac{\ell_t}{24}}$$

is a weight-1/2 weakly holomorphic modular form on $\Gamma_1(144 f_t^2 \widetilde{\ell}_t \mathcal{P}^4)$.

Sketch of the proof. The proof requires the Fourier expansions of the forms $D\left(\frac{a}{c}; z\right)$. To give these expansions, we require the incomplete gamma function

$$\Gamma(a; x) := \int_x^\infty e^{-t} t^{a-1}\, dt. \tag{3.20}$$

For integers $0 < a < c$, we have

$$D\left(\frac{a}{c}; z\right) = q^{-\frac{\ell_c}{24}} + \sum_{n=1}^\infty \sum_{m=-\infty}^\infty N(m, n) \zeta_c^{am} q^{\ell_c n - \frac{\ell_c}{24}}$$

$$+ \frac{i \sin\left(\frac{\pi a}{c}\right) \ell_c^{\frac{1}{2}}}{\sqrt{3}} \sum_{m \,(\mathrm{mod}\ f_c)} (-1)^m \sin\left(\frac{a\pi(6m+1)}{c}\right)$$

$$\times \sum_{n \equiv 6m+1 \,(\mathrm{mod}\ 6f_c)} \gamma(c, y; n) q^{-\widetilde{\ell}_c n^2}, \tag{3.21}$$

where

$$\gamma(c, y; n) := \frac{i}{\sqrt{2\pi \ell_c}} \cdot \Gamma\left(\frac{1}{2}; 4\pi \widetilde{\ell}_c n^2 y\right).$$

This expansion follows easily from

$$-S_1\left(\frac{a}{c}; z\right) = \frac{i \sin\left(\frac{\pi a}{c}\right) \ell_c^{\frac{1}{2}}}{\sqrt{3}} \sum_{m \,(\mathrm{mod}\ f_c)} (-1)^m \sin\left(\frac{a\pi(6m+1)}{c}\right)$$

$$\times \sum_{n \equiv 6m+1 \,(\mathrm{mod}\ 6f_c)} \int_{-\bar{z}}^{i\infty} \frac{n e^{2\pi i n^2 \widetilde{\ell}_c \tau}}{\sqrt{-i(\tau + z)}}\, d\tau$$

and the integral identity

$$\int_{-\bar{z}}^{i\infty} \frac{n e^{2\pi i n^2 \widetilde{\ell}_c \tau}}{\sqrt{-i(\tau + z)}}\, d\tau = \gamma(c, y; n) \cdot q^{-\widetilde{\ell}_c n^2}.$$

The key point of the proof is that the nonholomorphic parts of these weak Maass forms have the property that their coefficients are supported on a fixed square class, one that is easily annihilated by taking linear combinations of quadratic twists. In particular, suppose that $\mathcal{P} \nmid 6c$ is prime. For this prime $\mathcal{P}$, let

$$g := \sum_{v=1}^{\mathcal{P}-1} \left(\frac{v}{\mathcal{P}}\right) e^{\frac{2\pi i v}{\mathcal{P}}}$$

be the usual Gauss sum with respect to $\mathcal{P}$. Define the function $D\left(\frac{a}{c}; z\right)_{\mathcal{P}}$ by

$$D\left(\frac{a}{c}; z\right)_{\mathcal{P}} := \frac{g}{\mathcal{P}} \sum_{v=1}^{\mathcal{P}-1} \left(\frac{v}{\mathcal{P}}\right) D\left(\frac{a}{c}; z\right) \Big|_{\frac{1}{2}} \begin{pmatrix} 1 & -\frac{v}{\mathcal{P}} \\ 0 & 1 \end{pmatrix}, \tag{3.22}$$

where $|_{\frac{1}{2}}$ is the usual "slash operator" (for example, see page 51 of [26]). By construction, $D\left(\frac{a}{c};z\right)_{\mathcal{P}}$ is the $\mathcal{P}$ quadratic twist of $D\left(\frac{a}{c};z\right)$. In other words, the nth coefficient in the q-expansion of $D\left(\frac{a}{c};z\right)_{\mathcal{P}}$ is $\left(\frac{n}{\mathcal{P}}\right)$ times the nth coefficient of $D\left(\frac{a}{c};z\right)$. For the nonholomorphic part, this follows from the fact that the factors $\gamma(c,y;n)$ appearing in (3.21) are fixed by the transformations in (3.22).

Generalizing classical facts about twists of modular forms, $D\left(\frac{a}{c};z\right)_{\mathcal{P}}$ is a weak Maass form of weight $1/2$ on $\Gamma_1(144 f_c^2 \widetilde{\ell}_c \mathcal{P}^2)$. By (3.21), it follows that

$$D\left(\frac{a}{c};z\right) - \left(\frac{-\widetilde{\ell}_c}{\mathcal{P}}\right) D\left(\frac{a}{c};z\right)_{\mathcal{P}} \tag{3.23}$$

is a weak Maass form of weight $1/2$ on $\Gamma_1(144 f_c^2 \widetilde{\ell}_c \mathcal{P}^2)$ with the property that its nonholomorphic part is supported on summands of the form $*q^{-\widetilde{\ell}_c \mathcal{P}^2 n^2}$. These terms are annihilated by taking the $\mathcal{P}$-quadratic twist of this Maass form. Consequently, we obtain a weakly holomorphic modular form of weight $1/2$ on $\Gamma_1(144 f_c^2 \widetilde{\ell}_c \mathcal{P}^4)$. Thanks to (3.19), when $t = c$, the conclusion of Theorem 3.5 follows. $\qquad \square$

Theorem 3.5 allows us to employ the rich theory of weakly holomorphic modular forms in the study of partition ranks. Here we describe results that were originally inspired by the celebrated Ramanujan congruences:

$$p(5n + 4) \equiv 0 \pmod{5},$$

$$p(7n + 5) \equiv 0 \pmod{7},$$

$$p(11n + 6) \equiv 0 \pmod{11}.$$

In his famous seminal paper [17], Dyson conjectured that ranks could be used to provide a combinatorial "explanation" for the first two congruences.[1] Precisely, he conjectured[2] that for every integer n and every r we have

$$N(r, 5; 5n + 4) = \frac{p(5n + 4)}{5}, \tag{3.24}$$

$$N(r, 7; 7n + 5) = \frac{p(7n + 5)}{7}. \tag{3.25}$$

In an important paper, Atkin and Swinnerton-Dyer [9] confirmed Dyson's conjecture in 1954. It is not difficult to use Theorem 3.1 to give alternative proofs of these rank identities, as well as others of similar type.

---

[1] He further postulated the existence of another statistic, the so-called 'crank' that could be used to provide an explanation for all three Ramanujan congruences. In 1988, Andrews and Garvan [8] found the crank, and they confirmed Dyson's speculation that it explains the three Ramanujan congruences. Recent work of Mahlburg [24] establishes that the Andrews–Dyson–Garvan crank plays an even more central role in the theory of partition congruences. His work establishes congruences modulo arbitrary powers of all primes ≥ 5. Other work by Garvan, Kim, and Stanton [19] gives a different "crank" for several other Ramanujan congruences.

[2] A short calculation reveals that this phenomenon cannot hold modulo 11.

Although identities such as (3.24) and (3.25) are rare, it is still natural to use Theorem 3.5 to investigate the relation between ranks and generic partition congruences such as

$$p(48037937n + 1122838) \equiv 0 \pmod{17},$$

$$p(1977147619n + 815655) \equiv 0 \pmod{19},$$

$$p(14375n + 3474) \equiv 0 \pmod{23},$$

$$p(348104768909n + 43819835) \equiv 0 \pmod{29},$$

$$p(4063467631n + 30064597) \equiv 0 \pmod{31},$$

which are now known to exist (for example, see [10, 25, 2, 1]). We shall employ a method first used by the author in [25] in his work on $p(n)$, and we show that Dyson's rank-partition functions themselves uniformly satisfy many congruences of Ramanujan type.

Theorem 3.6 (Bringmann–Ono, Theorem 1.5 of [12]). *Let t be a positive odd integer, and let $Q \nmid 6t$ be prime. If j is a positive integer, then there are infinitely many nonnested arithmetic progressions $An + B$ such that for every $0 \le r < t$ we have*

$$N(r, t; An + B) \equiv 0 \pmod{Q^j}.$$

Two remarks.
(1) Theorem 3.6 provides a combinatorial decomposition of the partition function congruence

$$p(An + B) \equiv 0 \pmod{Q^j}.$$

(2) By "nonnested," we mean that there are infinitely many arithmetic progressions $An + B$, with $0 \le B < A$, with the property that there are no progressions that contain another progression.

Unlike Dyson's original conjectures for the congruences with modulus 5 and 7, and the work of Mahlburg [24], Theorem 3.6 says nothing about those primes $Q \ge 5$ that may divide t. It is nearly certain that this is a consequence of a nonoptimal proof. There is good reason to suspect the truth of the following conjecture.

Conjecture. Theorem 3.6 holds for those primes $Q \ge 5$ that divide t.

Sketch of the proof of Theorem 3.6. The proof depends on Theorem 3.5, the observation that certain "sieved" partition rank-generating functions are weakly holomorphic modular forms. In short, this result reduces the proof of Theorem 3.6 to the fact that any finite number of half-integral-weight cusp forms with integer coefficients are annihilated, modulo a fixed prime power, by a positive proportion of half-integral-weight Hecke operators.

To be precise, suppose that $f_1(z), f_2(z), \ldots, f_s(z)$ are half-integral-weight cusp forms, where

$$f_i(z) \in S_{\lambda_i + \frac{1}{2}}(\Gamma_1(4N_i)) \cap \mathcal{O}_K[[q]],$$

and where $\mathcal{O}_K$ is the ring of integers of a fixed number field K. If Q is prime and $j \geq 1$ is an integer, then the set of primes L for which

$$f_i(z) \mid T_{\lambda_i}(L^2) \equiv 0 \pmod{Q^j}, \tag{3.26}$$

for each $1 \leq i \leq s$, has positive Frobenius density. Here $T_{\lambda_i}(L^2)$ denotes the usual L^2 index Hecke operator of weight $\lambda_i + \frac{1}{2}$.

Suppose that $\mathcal{P} \nmid 6tQ$ is prime. By Theorem 3.5, for every $0 \leq r < t$,

$$F(r, t, \mathcal{P}; z) = \sum_{n=1}^{\infty} a(r, t, \mathcal{P}; n)q^n$$

$$:= \sum_{\left(\frac{24\ell_t n - \ell_t}{\mathcal{P}}\right) = -\left(\frac{-24\widetilde{\ell_t}}{\mathcal{P}}\right)} \left(N(r, t; n) - \frac{p(n)}{t} \right) q^{\ell_t n - \frac{\ell_t}{24}} \tag{3.27}$$

is a weakly holomorphic modular form of weight $1/2$ on $\Gamma_1(144 f_t^2 \widetilde{\ell_t} \mathcal{P}^4)$. Furthermore, by the work of Ahlgren and the author [2], it follows that

$$P(t, \mathcal{P}; z) = \sum_{n=1}^{\infty} p(t, \mathcal{P}; n)q^n := \sum_{\left(\frac{24\ell_t n - \ell_t}{\mathcal{P}}\right) = -\left(\frac{-24\widetilde{\ell_t}}{\mathcal{P}}\right)} p(n)q^{\ell_t n - \frac{\ell_t}{24}} \tag{3.28}$$

is a weakly holomorphic modular form of weight $-1/2$ on $\Gamma_1(576 \widetilde{\ell_t} \mathcal{P}^4)$. In particular, all of these forms are modular with respect to $\Gamma_1(576 f_t^2 \widetilde{\ell_t} \mathcal{P}^4)$.

Since $Q \nmid 576 f_t^2 \widetilde{\ell_t} \mathcal{P}^4$, a result of Treneer (see Theorem 3.1 of [32]), generalizing earlier observations of Ahlgren and Ono [2, 25], implies that there is a sufficiently large integer m for which

$$\sum_{Q \nmid n} a(r, t, \mathcal{P}; Q^m n)q^n,$$

for all $0 \leq r < t$, and

$$\sum_{Q \nmid n} p(t, \mathcal{P}; Q^m n)q^n$$

are all congruent modulo Q^j to forms in the graded ring of half-integral-weight cusp forms with algebraic integer coefficients on $\Gamma_1(576 f_t^2 \widetilde{\ell_t} \mathcal{P}^4 Q^2)$.

The system of simultaneous congruences (3.26), in the case of these forms, guarantees that a positive proportion of primes L have the property that these forms modulo Q^j are annihilated by the index-L^2 half-integral-weight Hecke operators. Theorem 3.6 now follows mutatis mutandis as in the proof of Theorem 1 of [25]. $\square$

Two remarks.

(1) The simultaneous system (3.26) of congruences follows from a straightforward generalization of a classical observation of Serre (see Section 6 of [30]).

(2) Treneer states her result for weakly holomorphic modular forms on $\Gamma_0(4N)$ with Nebentypus. We are using a straightforward extension of her result to $\Gamma_1(4N)$.

3.3 The Andrews–Dragonette Conjecture for $f(q)$

Rademacher famously employed the modularity of (1.1) to perfect the Hardy–Ramanujan asymptotic formula

$$p(n) \sim \frac{1}{4n\sqrt{3}} \cdot e^{\pi \sqrt{2n/3}} \tag{3.29}$$

to obtain his exact formula for $p(n)$. To state his formula, let $I_s(x)$ be the usual I-Bessel function of order s. Furthermore, if $k \geq 1$ and n are integers, then let

$$A_k(n) := \frac{1}{2}\sqrt{\frac{k}{12}} \sum_{\substack{x \pmod{24k} \\ x^2 \equiv -24n+1 \pmod{24k}}} \chi_{12}(x) \cdot e\left(\frac{x}{12k}\right), \tag{3.30}$$

where the sum runs over the residue classes modulo $24k$, and where

$$\chi_{12}(x) := \left(\frac{12}{x}\right). \tag{3.31}$$

If n is a positive integer, then one version of Rademacher's formula reads [27]

$$p(n) = 2\pi(24n - 1)^{-\frac{3}{4}} \sum_{k=1}^{\infty} \frac{A_k(n)}{k} \cdot I_{\frac{3}{2}}\left(\frac{\pi \sqrt{24n - 1}}{6k}\right). \tag{3.32}$$

If $N_e(n)$ (respectively $N_o(n)$) denotes the number of partitions of n with even (respectively odd) rank, then by letting $w = -1$ in (3.2) we obtain

$$1 + \sum_{n=1}^{\infty}(N_e(n) - N_o(n))q^n = 1 + \sum_{n=1}^{\infty} \frac{q^{n^2}}{(1+q)^2(1+q^2)^2 \cdots (1+q^n)^2}. \tag{3.33}$$

In the spirit of Rademacher's work, it is natural to seek exact formulas for $N_e(n)$ and $N_o(n)$. In view of (3.32) and (3.33), since

$$p(n) = N_e(n) + N_o(n),$$

this question is equivalent to the problem of deriving exact formulas for the coefficients $\alpha(n)$ of the mock theta function

$$f(q) = 1 + \sum_{n=1}^{\infty} \alpha(n)q^n := 1 + \sum_{n=1}^{\infty} \frac{q^{n^2}}{(1+q)^2(1+q^2)^2 \cdots (1+q^n)^2}. \tag{3.34}$$

The problem of estimating the coefficients $\alpha(n)$ has a long history, one that even precedes Dyson's definition of partition ranks. Indeed, Ramanujan's last letter to Hardy already includes the claim that

$$\alpha(n) = (-1)^{n-1} \frac{\exp\left(\pi\sqrt{\frac{n}{6} - \frac{1}{144}}\right)}{2\sqrt{n - \frac{1}{24}}} + O\left(\frac{\exp\left(\frac{1}{2}\pi\sqrt{\frac{n}{6} - \frac{1}{144}}\right)}{\sqrt{n - \frac{1}{24}}}\right).$$

Typical of his writings, Ramanujan offered no proof of this claim. Dragonette proved this claim in her 1951 Ph.D. thesis [16], and Andrews [3] subsequently improved upon Dragonette's work, and he proved[3] that

$$\alpha(n) = \pi(24n - 1)^{-\frac{1}{4}} \sum_{k=1}^{\lfloor\sqrt{n}\rfloor} \frac{(-1)^{\lfloor\frac{k+1}{2}\rfloor} A_{2k}\left(n - \frac{k(1+(-1)^k)}{4}\right)}{k} I_{\frac{1}{2}}\left(\frac{\pi\sqrt{24n-1}}{12k}\right)$$

$$+ O(n^\epsilon). \tag{3.35}$$

This result falls short of the problem of obtaining an exact formula for $\alpha(n)$. In his plenary address "Partitions: At the interface of q-series and modular forms," delivered at the Millennial Number Theory Conference at the University of Illinois in 2000, Andrews highlighted this classical problem by promoting his conjecture of 1966 (see page 456 of [3], and Section 5 of [5]) for the coefficients $\alpha(n)$.

Conjecture (Andrews–Dragonette). If n is a positive integer, then

$$\alpha(n) = \pi(24n - 1)^{-\frac{1}{4}} \sum_{k=1}^{\infty} \frac{(-1)^{\lfloor\frac{k+1}{2}\rfloor} A_{2k}\left(n - \frac{k(1+(-1)^k)}{4}\right)}{k} \cdot I_{\frac{1}{2}}\left(\frac{\pi\sqrt{24n-1}}{12k}\right).$$

$$\tag{3.36}$$

Bringmann and the author have proved [11] the following theorem.

Theorem 3.7 (Bringmann–Ono, Theorem 1.1 of [11]). *The Andrews–Dragonette conjecture is true.*

Sketch of the proof. By a more precise version of Theorem 3.1, which is easily deduced from work of Zwegers [35], we have that $D(\frac{1}{2}; z)$ is a weight 1/2 weak Maass form on $\Gamma_0(144)$ with Nebentypus character $\chi_{12} = \left(\frac{12}{\bullet}\right)$. The idea behind the proof is simple. We shall construct a Maass–Poincaré series that we shall show equals $D(\frac{1}{2}; z)$. The proof of the conjecture then follows from the fact that the formulas in the Andrews–Dragonette conjecture can be shown to give the coefficients of this Maass–Poincaré series.

Suppose that $k \in \frac{1}{2} + \mathbb{Z}$. We define a class of Poincaré series $P_k(s; z)$. For matrices $\begin{pmatrix} a & b \\ c & d \end{pmatrix} \in \Gamma_0(2)$, with $c \geq 0$, define the character $\chi(\cdot)$ by

$$\chi\left(\begin{pmatrix} a & b \\ c & d \end{pmatrix}\right) := \begin{cases} e\left(-\frac{b}{24}\right) & \text{if } c = 0, \\ i^{-1/2}(-1)^{\frac{1}{2}(c+ad+1)} e\left(-\frac{a+d}{24c} - \frac{a}{4} + \frac{3dc}{8}\right) \cdot \omega_{-d,c}^{-1} & \text{if } c > 0. \end{cases}$$

$$\tag{3.37}$$

---

[3] This is a reformulation of Theorem 5.1 of [3] using the identity $I_{\frac{1}{2}}(z) = \left(\frac{2}{\pi z}\right)^{\frac{1}{2}} \cdot \sinh(z)$.

Throughout, let $z = x + iy$, and for $s \in \mathbb{C}$, $k \in \frac{1}{2} + \mathbb{Z}$, and $y \in \mathbb{R} \setminus \{0\}$, let

$$\mathcal{M}_s(y) := |y|^{-\frac{k}{2}} M_{\frac{k}{2} \operatorname{sgn}(y), \, s-\frac{1}{2}}(|y|), \tag{3.38}$$

where $M_{\nu,\mu}(z)$ is the standard M-Whittaker function, which is a solution to the differential equation

$$\frac{\partial^2 u}{\partial z^2} + \left(-\frac{1}{4} + \frac{\nu}{z} + \frac{\frac{1}{4} - \mu^2}{z^2} \right) u = 0. \tag{3.39}$$

Furthermore, let

$$\varphi_{s,k}(z) := \mathcal{M}_s\left(-\frac{\pi y}{6} \right) e\left(-\frac{x}{24} \right).$$

Using this notation, define the Poincaré series $P_k(s; z)$ by

$$P_k(s; z) := \frac{2}{\sqrt{\pi}} \sum_{M \in \Gamma_\infty \backslash \Gamma_0(2)} \chi(M)^{-1} (cz + d)^{-k} \varphi_{s,k}(Mz). \tag{3.40}$$

Here Γ_∞ is the subgroup of translations in $\mathrm{SL}_2(\mathbb{Z})$:

$$\Gamma_\infty := \left\{ \pm \begin{pmatrix} 1 & n \\ 0 & 1 \end{pmatrix} : n \in \mathbb{Z} \right\}.$$

The defining series is absolutely convergent for $P_k\left(1 - \frac{k}{2}; z \right)$ for $k < 1/2$, and is conditionally convergent when $k = 1/2$. We are interested in $P_{\frac{1}{2}}\left(\frac{3}{4}; z \right)$, which we define by analytically continuing Fourier expansions. This argument is not straight-forward (see Theorem 3.2 and Corollary 4.2 of [11]). Thanks to (3.39), as a result we find that $P_{\frac{1}{2}}\left(\frac{3}{4}; 24z \right)$ is a weak Maass form of weight $1/2$ for $\Gamma_0(144)$ with Nebentypus χ_{12}.

After a long calculation, one can show that this Maass–Poincaré series has the Fourier expansion

$$P_{\frac{1}{2}}\left(\frac{3}{4}; z \right)$$

$$= \left(1 - \pi^{-\frac{1}{2}} \cdot \Gamma\left(\frac{1}{2}, \frac{\pi y}{6} \right) \right) \cdot q^{-\frac{1}{24}} + \sum_{n=-\infty}^{0} \gamma_y(n) q^{n-\frac{1}{24}} + \sum_{n=1}^{\infty} \beta(n) q^{n-\frac{1}{24}}, \tag{3.41}$$

where for positive integers n we have

$$\beta(n) = \pi (24n - 1)^{-\frac{1}{4}} \sum_{k=1}^{\infty} \frac{(-1)^{\lfloor \frac{k+1}{2} \rfloor} A_{2k}\left(n - \frac{k(1+(-1)^k)}{4} \right)}{k} \cdot I_{\frac{1}{2}}\left(\frac{\pi \sqrt{24n - 1}}{12k} \right), \tag{3.42}$$

and for nonpositive integers n we have

$$\gamma_y(n) = \pi^{\frac{1}{2}}|24n - 1|^{-\frac{1}{4}} \cdot \Gamma\left(\frac{1}{2}, \frac{\pi|24n - 1| \cdot y}{6}\right)$$

$$\times \sum_{k=1}^{\infty} \frac{(-1)^{\lfloor\frac{k+1}{2}\rfloor} A_{2k}\left(n - \frac{k(1+(-1)^k)}{4}\right)}{k} \cdot J_{\frac{1}{2}}\left(\frac{\pi\sqrt{|24n - 1|}}{12k}\right).$$

For convenience, we let

$$P(z) := P_{\frac{1}{2}}\left(\frac{3}{4}; 24z\right). \tag{3.43}$$

Canonically decompose $P(z)$ into a nonholomorphic and a holomorphic part:

$$P(z) = P_{nh}(z) + P_h(z). \tag{3.44}$$

In particular, we have that

$$P_h(z) = q^{-1} + \sum_{n=1}^{\infty} \beta(n)q^{24n-1}.$$

Since $P(z)$ and $D\left(\frac{1}{2}; z\right)$ are weak Maass forms of weight $1/2$ for $\Gamma_0(144)$ with Nebentypus χ_{12}, (3.41) and (3.42) imply that the proof of the conjecture reduces to proving that these forms are equal. First, one shows that

$$P_{nh}(z) = -S_1\left(\frac{1}{2}; z\right). \tag{3.45}$$

To establish this, we apply Proposition 2.1. One can show that $\xi_{\frac{1}{2}}(P(z))$ is a holomorphic modular form of weight $3/2$ on $\Gamma_0(144)$ with Nebentypus χ_{12}. Using (3.41), it can be shown that the nonzero coefficients of $\xi_{\frac{1}{2}}(P(z))$ are supported on exponents in the arithmetic progression 1 (mod 24). Now we apply $\xi_{\frac{1}{2}}$ to $D\left(\frac{1}{2}; z\right)$, and we find that

$$\xi_{\frac{1}{2}}\left(D\left(\frac{1}{2}; z\right)\right) = 4\vartheta(\psi; z) = 4\sum_{n=1}^{\infty} \psi(n)\, n\, q^{n^2},$$

where

$$\psi(n) := \begin{cases} -1 & \text{if } n \equiv 1 \pmod 6, \\ -1 & \text{if } n \equiv 5 \pmod 6. \end{cases}$$

Obviously, its Fourier coefficients are also supported on exponents in the arithmetic progression 1 (mod 24). Therefore, $\xi_{\frac{1}{2}}(P(z))$ and $\xi_{\frac{1}{2}}\left(D\left(\frac{1}{2}; z\right)\right)$ are both holomorphic modular forms of weight $3/2$ on $\Gamma_0(144)$ with Nebentypus χ_{12}. Using dimension formulas for spaces of half-integral-weight modular forms and the Serre–Stark basis theorem, it follows that

$$\dim_{\mathbb{C}}(M_{1/2}(\Gamma_0(144), \chi_{12})) = \dim_{\mathbb{C}}(S_{1/2}(\Gamma_0(144), \chi_{12})) = 0 \tag{3.46}$$

implies that

$$\dim_{\mathbb{C}} \left(M_{3/2} \left(\Gamma_0(144), \chi_{12} \right) \right) = 24.$$

Since $\xi_{\frac{1}{2}}(P(z)), \xi_{\frac{1}{2}}\left(D\left(\frac{1}{2}; z\right)\right) \in M_{3/2}(\Gamma_0(144), \chi_{12})$ both have the property that their Fourier coefficients are supported on exponents of the form $24n + 1 \geq 1$, choose a constant c such that the coefficient of q, and hence all the coefficients up to and including q^{24}, of $\xi_{\frac{1}{2}}(P(z))$ and $c \cdot \xi_{\frac{1}{2}}\left(D\left(\frac{1}{2}; z\right)\right)$ agree. By dimensionality, this in turn implies that $\xi_{\frac{1}{2}}(P(z)) = c \cdot \xi_{\frac{1}{2}}\left(D\left(\frac{1}{2}; z\right)\right)$, and so we have that

$$P_{nh}(z) = -c \cdot S_1 \left(\frac{1}{2}; z \right).$$

To establish that $c = 1$, let

$$E(z) := P(z) - c \cdot D \left(\frac{1}{2}; z \right).$$

This function is a weakly holomorphic modular form of weight $1/2$ on $\Gamma_0(144)$ with Nebentypus χ_{12}. By (3.35) and Corollary 4.2 of [11], it follows that

$$E(z) = P_h(z) - c \cdot D \left(\frac{1}{2}; z \right) = (1 - c)q^{-1}f(q^{24}) + \sum_{n \geq 0} A(n)q^{24n-1},$$

where $|A(n)| = O\left((24n - 1)^{\frac{3}{4}+\epsilon}\right)$ for positive integers n. By work of Zwegers [35], applying the map $z \mapsto -\frac{1}{z}$ returns a nonholomorphic contribution unless $c = 1$. Since $E(z)$ does not have a nonholomorphic component, it follows that $c = 1$, which in turn proves that $P_{nh}(z) = -S_1\left(\frac{1}{2}; z\right)$.

Hence it follows that

$$P(z) - D \left(\frac{1}{2}; z \right) = P_h(z) - q^{-1}R(-1; q^{24})$$

$$= q^{-1} + \sum_{n=1}^{\infty} \beta(n)q^{24n-1} - q^{-1}f(q^{24}) = \sum_{n=1}^{\infty} \nu(n)q^{24n-1}$$

is a weakly holomorphic modular form of weight $1/2$ on $\Gamma_0(144)$ with Nebentypus χ_{12}. By (3.35) and Corollary 4.2 of [11] again, it can be shown that

$$|\nu(n)| = O \left(n^{\frac{3}{4}+\epsilon} \right).$$

Therefore, $P(z) - D\left(\frac{1}{2}; z\right)$ is a holomorphic modular form. However, by (3.46), this space is trivial, and so we find that $P(z) - D\left(\frac{1}{2}; z\right) = 0$, which completes the proof.

$\square$

References

1. S. Ahlgren, *Distribution of the partition function modulo composite integers M*, Math. Annalen, **318** (2000), pages 795–803.
2. S. Ahlgren and K. Ono, *Congruence properties for the partition function*, Proc. Natl. Acad. Sci., USA **98**, No. 23 (2001), pages 12882–12884.
3. G. E. Andrews, *On the theorems of Watson and Dragonette for Ramanujan's mock theta functions*, Amer. J. Math. **88** No. 2 (1966), pages 454–490.
4. G. E. Andrews, *Mock theta functions,* Theta functions – Bowdoin 1987, Part 2 (Brunswick, ME., 1987), pages 283–297, Proc. Sympos. Pure Math. **49**, Part 2, Amer. Math. Soc., Providence, RI., 1989.
5. G. E. Andrews, *Partitions: At the interface of q-series and modular forms*, Rankin Memorial Issues, Ramanujan J. **7** (2003), pages 385–400.
6. G. E. Andrews, *Partitions with short sequences and mock theta functions*, Proc. Natl. Acad. Sci. USA, **102** No. 13 (2005), pages 4666–4671.
7. G. E. Andrews, F. Dyson, and D. Hickerson, *Partitions and indefinite quadratic forms*, Invent. Math. **91** No. 3 (1988), pages 391–407.
8. G. E. Andrews and F. Garvan, *Dyson's crank of a partition*, Bull. Amer. Math. Soc. (N.S.) **18** No. 2 (1988), pages 167–171.
9. A. O. L. Atkin and H. P. F. Swinnerton-Dyer, *Some properties of partitions*, Proc. London Math. Soc. **66** No. 4 (1954), pages 84–106.
10. A. O. L. Atkin, *Multiplicative congruence properties and density problems for p(n)*, Proc. London Math. Soc. **3** 18 (1968), pages 563–576.
11. K. Bringmann and K. Ono, *The f(q) mock theta function conjecture and partition ranks*, Invent. Math., **165** (2006), pages 243–266.
12. K. Bringmann and K. Ono, *Dyson's ranks and Maass forms*, Ann. of Math., accepted for publication.
13. J. H. Bruinier and J. Funke, *On two geometric theta lifts*, Duke Math. J. **125** (2004), pages 45–90.
14. Y.-S. Choi, *Tenth order mock theta functions in Ramanujan's lost notebook*, Invent. Math. **136** No. 3 (1999), pages 497–569.
15. H. Cohen, *q-identities for Maass waveforms*, Invent. Math. **91** No. 3 (1988), pages 409–422.
16. L. Dragonette, *Some asymptotic formulas for the mock theta series of Ramanujan*, Trans. Amer. Math. Soc. **72** No. 3 (1952), pages 474–500.
17. F. Dyson, *Some guesses in the theory of partitions*, Eureka (Cambridge) **8** (1944), pages 10–15.
18. F. Dyson, *A walk through Ramanujan's garden*, Ramanujan revisited (Urbana-Champaign, Ill. 1987), Academic Press, Boston, 1988, pages 7–28.
19. F. Garvan, D. Kim, and D. Stanton, *Cranks and t-cores*, Invent. Math. **101** (1990), pages 1–17.
20. B. Gordon and R. McIntosh, *Some eighth order mock theta functions,* J. London Math. Soc. **62** No. 2 (2000), pages 321–335.
21. B. Gordon and R. McIntosh, *Modular transformations of Ramanujan's fifth and seventh order mock theta functions*, Ramanujan J. **7** (2003), pages 193–222.
22. D. Hickerson, *On the seventh order mock theta functions*, Invent. Math. **94** No. 3 (1988), pages 661–677.
23. R. Lawrence and D. Zagier, *Modular forms and quantum invariants of 3-manifolds*, Asian J. Math. **3** (1999), pages 93–107.

24. K. Mahlburg, *Partition congruences and the Andrews–Garvan–Dyson crank*, Proc. Natl. Acad. Sci., USA, **102** (2005), pages 15373–15376.
25. K. Ono, *Distribution of the partition function modulo m*, Ann. of Math. **151** (2000), pages 293–307.
26. K. Ono, *The web of modularity: Arithmetic of the coefficients of modular forms and q-series*, Conf. Board of the Math. Sciences, Regional Conference Series, No. 102, Amer. Math. Soc., Providence, 2004.
27. H. Rademacher, *Topics in analytic number theory*, Die Grundlehren der mathematischen Wissenschaften, Band **169**, Springer-Verlag New York–Heidelberg, 1973.
28. S. Ramanujan, *The lost notebook and other unpublished papers*, Narosa, New Delhi, 1988.
29. A. Selberg, *Über die Mock-Thetafunktionen siebenter Ordnung*, Arch. Math. Naturviden- skab, **41** (1938), pages 3–15 (see also Coll. Papers, I, pages 22–37).
30. J.-P. Serre, *Divisibilité de certaines fonctions arithmétiques*, Enseign. Math. **22** (1976), pages 227–260.
31. G. Shimura, *On modular forms of half integral weight*, Ann. of Math. **97** (1973), pages 440–481.
32. S. Treneer, *Congruences for the coefficients of weakly holomorphic modular forms*, Proc. London Math. Soc., **93** (2006), pages 304–324.
33. G. N. Watson, *The final problem: An account of the mock theta functions*, J. London Math. Soc. **2** (2) (1936), pages 55–80.
34. G. N. Watson, *The mock theta functions (2)*, Proc. London Math. Soc. (2) **42** (1937), pages 274–304.
35. S. P. Zwegers, *Mock ϑ-functions and real analytic modular forms*, q-series with appli- cations to combinatorics, number theory, and physics (ed. B. C. Berndt and K. Ono), Contemp. Math. **291**, Amer. Math. Soc., (2001), pages 269–277.
36. S. P. Zwegers, *Mock theta functions*, Ph.D. Thesis, Universiteit Utrecht, 2002.

7

Elliptic Functions and Transcendence

Michel Waldschmidt

Université P. et M. Curie (Paris VI), Institut de Mathématiques de Jussieu,
UMR 7586 CNRS, Problèmes Diophantiens, Case 247, 175, rue du Chevaleret F-75013
Paris, France
miw@math.jussieu.fr; http://www.math.jussieu.fr/~miw/

Summary. Transcendental numbers form a fascinating subject: so little is known about the nature of analytic constants that more research is needed in this area. Even when one is interested only in numbers like π and e^π that are related to the classical exponential function, it turns out that elliptic functions are required (so far, this should not last forever!) to prove transcendence results and get a better understanding of the situation.

First we briefly recall some of the basic transcendence results related to the exponential function (Section 1). Next, in Section 2, we survey the main properties of elliptic functions that are involved in transcendence theory.

We survey transcendence theory of values of elliptic functions in Section 3, linear independence in Section 4, and algebraic independence in Section 5. This splitting is somewhat artificial but convenient. Moreover, we restrict ourselves to elliptic functions, even when many results are only special cases of statements valid for abelian functions. A number of related topics are not considered here (e.g., heights, p-adic theory, theta functions, Diophantine geometry on elliptic curves).

Key words: Transcendental numbers, elliptic functions, elliptic curves, elliptic integrals, algebraic independence, transcendence measures, measures of algebraic independence, Diophantine approximation.

2000 *Mathematics Subject Classifications*: 01-02, 11G05, 11J89

Mathematical history lecture given on February 21, 2005, at the Mathematics Department of the University of Florida for the *Special Year in Number Theory and Combinatorics 2004–05*, supported by the France-Florida Research Institute (FFRI). The author is thankful to Krishna Alladi for his invitation to deliver the lecture and for his suggestion of coming up with a written version. He is also grateful to Nikos Tzanakis for his proposal to deliver a course for the Pichorides Distinguished Lectureship funded by FORTH (Foundation of Research and Technology Hellas) during the summer 2005 at the Department of Mathematics, University of Crete, where this survey was written. Last but not least, many thanks to F. Amoroso, D. Bertrand, N. Brisebarre, P. Bundschuh, E. Gaudron, N. Hirata-Kohno, C. Levesque, D. W. Masser, F. & R. Pellarin, P. Philippon, I. Wakabayashi, and all colleagues who contributed their helpful comments on preliminary versions of this survey. Université P. et M. Curie (Paris VI). http://www.math.jussieu.fr/~miw/

K. Alladi (ed.), *Surveys in Number Theory*, DOI: 10.1007/978-0-387-78510-3_7,

1 Exponential Function and Transcendence

We start with a very brief list of some of the main transcendence results concerning numbers related to the exponential function. References are, for instance, [13, 83, 89, 120, 192, 207, 211, 233].

Next, we point out some properties of the exponential function, the elliptic analogue of which we shall consider later (Section 2.1).

1.1 Short Survey on the Transcendence of Numbers Related to the Exponential Function

Hermite, Lindemann, and Weierstrass

The first transcendence result for a number related to the exponential function is Hermite's theorem on the transcendence of e.

Theorem 1 (Hermite, 1873). *The number e is transcendental.*

This means that for any nonzero polynomial $P \in \mathbb{Z}[X]$, the number $P(e)$ is not zero. We denote by $\overline{\mathbb{Q}}$ the set of algebraic numbers. Hence Hermite's theorem can be written $e \notin \overline{\mathbb{Q}}$. A complex number is called *transcendental* if it is transcendental over $\mathbb{Q}$, or over $\overline{\mathbb{Q}}$, which is the same. Also we shall say that complex numbers $\theta_1, \ldots, \theta_n$ are *algebraically independent* if they are algebraically independent over $\mathbb{Q}$, which is the same as over $\overline{\mathbb{Q}}$: for any nonzero polynomial P in n variables (and coefficients in $\mathbb{Z}$, $\mathbb{Q}$, or $\overline{\mathbb{Q}}$), the number $P(\theta_1, \ldots, \theta_n)$ is not zero.

The second result in chronological order is Lindemann's theorem on the transcendence of π.

Theorem 2 (Lindemann, 1882). *The number π is transcendental.*

The next result contains the transcendence of both numbers e and π:

Theorem 3 (Hermite–Lindemann, 1882). *For $\alpha \in \overline{\mathbb{Q}}^\times$, any nonzero logarithm $\log \alpha$ of α is transcendental.*

We denote by $\mathcal{L}$ the $\mathbb{Q}$-vector space of logarithms of algebraic numbers:

$$\mathcal{L} = \left\{ \log \alpha \; ; \; \alpha \in \overline{\mathbb{Q}}^\times \right\} = \left\{ \ell \in \mathbb{C} \; ; \; e^\ell \in \overline{\mathbb{Q}}^\times \right\} = \exp^{-1}(\overline{\mathbb{Q}}^\times).$$

Hence Theorem 3 means that $\mathcal{L} \cap \overline{\mathbb{Q}} = \{0\}$. An alternative form is the following:

Theorem 4 (Hermite–Lindemann, 1882). *For any $\beta \in \overline{\mathbb{Q}}^\times$, the number e^β is transcendental.*

The first result of algebraic independence for the values of the exponential function goes back to the end of the nineteenth century.

Theorem 5 (Lindemann–Weierstrass, 1885). *Let* $\beta_1, \ldots, \beta_n$ *be* $\mathbb{Q}$*-linearly independent algebraic numbers. Then the numbers* $e^{\beta_1}, \ldots, e^{\beta_n}$ *are algebraically independent.*

Again, there is an alternative form of Theorem 5: it amounts to a statement of linear independence.

Theorem 6 (Lindemann–Weierstrass, 1885). *Let* $\gamma_1, \ldots, \gamma_m$ *be distinct algebraic numbers. Then the numbers* $e^{\gamma_1}, \ldots, e^{\gamma_m}$ *are linearly independent over* $\overline{\mathbb{Q}}$.

It is not difficult to check that Theorem 6 is equivalent to Theorem 5 with the conclusion that $e^{\beta_1}, \ldots, e^{\beta_n}$ are algebraically independent over $\overline{\mathbb{Q}}$; since it is equivalent to saying that $e^{\beta_1}, \ldots, e^{\beta_n}$ are algebraically independent over $\mathbb{Q}$, one does not lose anything if one changes the conclusion of Theorem 6 by stating that the numbers $e^{\gamma_1}, \ldots, e^{\gamma_m}$ are linearly independent over $\mathbb{Q}$.

Hilbert's Seventh Problem, Gel'fond and Schneider

The solution of Hilbert's seventh problem on the transcendence of α^β was obtained by Gel'fond and Schneider in 1934 (see [89, 207]).

Theorem 7 (Gel'fond–Schneider, 1934). *For* α *and* β *algebraic numbers with* $\alpha \neq 0$ *and* $\beta \notin \mathbb{Q}$ *and for any choice of* $\log \alpha \neq 0$, *the number* $\alpha^\beta = \exp(\beta \log \alpha)$ *is transcendental.*

This means that the two algebraically independent functions e^z and $e^{\beta z}$ cannot take algebraic values at the points $\log \alpha$ (A.O. Gel'fond) and also that the two algebraically independent functions z and $\alpha^z = e^{z \log \alpha}$ cannot take algebraic values at the points $m + n\beta$ with $(m, n) \in \mathbb{Z}^2$ (Th. Schneider).

Examples (quoted by D. Hilbert in 1900) of numbers whose transcendence follows from Theorem 7 are $2^{\sqrt{2}}$ and e^π (recall that $e^{i\pi} = -1$). The transcendence of e^π had already been proved in 1929 by A.O. Gel'fond.

Here is an equivalent statement to Theorem 7:

Theorem 8 (Gel'fond–Schneider, 1934). *Let* $\log \alpha_1, \log \alpha_2$ *be two nonzero logarithms of algebraic numbers. Assume that the quotient* $(\log \alpha_1)/(\log \alpha_2)$ *is irrational. Then this quotient is transcendental.*

Linear Independence of Logarithms of Algebraic Numbers

The generalization of Theorem 8 to more than two logarithms, conjectured by A.O. Gel'fond [89], was proved by A. Baker in 1966. His results include not only Theorem 8 but also Theorem 3.

Theorem 9 (Baker, 1966). *Let* $\log \alpha_1, \ldots, \log \alpha_n$ *be* $\mathbb{Q}$*-linearly independent logarithms of algebraic numbers. Then the numbers* $1, \log \alpha_1, \ldots, \log \alpha_n$ *are linearly independent over the field* $\overline{\mathbb{Q}}$.

The Six Exponentials Theorem and the Four Exponentials Conjecture

The next result, which does not follow from any of the previously mentioned results, was proved independently in the 1940s by C.L. Siegel (unpublished) and in the 1960s by S. Lang and K. Ramachandra (see [120, 191, 238]; see also Problem 1 in [207] for the four exponentials conjecture). As suggested by K. Ramachandra (see [192] Section 3.1, Theorem 2), Theorem 10 also follows from Schneider's criterion proved in 1949 [206].

Theorem 10 (Six Exponentials Theorem). *Let $x_1, \ldots, x_d$ be $\mathbb{Q}$-linearly indepen-dent complex numbers and let $y_1, \ldots, y_\ell$ be $\mathbb{Q}$-linearly independent complex numbers. Assume $\ell d > \ell + d$. Then at least one of the ℓd numbers*

$$e^{x_i y_j} \qquad (1 \le i \le d, \ 1 \le j \le \ell)$$

is transcendental.

Notice that the condition $\ell d > \ell + d$ can be written ($\ell \ge 2$ and $d \ge 3$) or ($\ell \ge 3$ and $d \ge 2$); it suffices to consider the case $\ell d = 6$ (hence the name of the result). Therefore, Theorem 10 can be stated in an equivalent form:

Theorem 11 (Six Exponentials Theorem—logarithmic form). *Let*

$$M = \begin{pmatrix} \log \alpha_1 & \log \alpha_2 & \log \alpha_3 \\ \log \beta_1 & \log \beta_2 & \log \beta_3 \end{pmatrix}$$

be a 2-by-3 matrix whose entries are logarithms of algebraic numbers. Assume that the three columns are linearly independent over $\mathbb{Q}$ and the two rows are also linearly independent over $\mathbb{Q}$. Then the matrix M has rank 2.

It is expected that the condition $d\ell > d + \ell$ in Theorem 10 is too restrictive and that the same conclusion holds in the case $d = \ell = 2$. We state this conjecture in the logarithmic form:

Conjecture 12 (Four exponentials conjecture — logarithmic form). Let

$$M = \begin{pmatrix} \log \alpha_1 & \log \alpha_2 \\ \log \beta_1 & \log \beta_2 \end{pmatrix}$$

be a 2-by-2 matrix whose entries are logarithms of algebraic numbers. Assume that the two columns are linearly independent over $\mathbb{Q}$ and that the two rows are also linearly independent over $\mathbb{Q}$. Then the matrix M has rank 2.

Algebraic Independence

In 1948 and 1949, A.O. Gel'fond extended his solution of Hilbert's seventh problem to a result of algebraic independence [89]. One of his theorems is that the two numbers $2^{\sqrt[3]{2}}$ and $2^{\sqrt[3]{4}}$ are algebraically independent. His general statements can be seen as extensions of Theorem 10 into a result of algebraic independence (in spite of the fact that Theorem 10 was stated and proved only several years later). In his original work, Gel'fond needed a stronger assumption, namely a measure of linear independence of the x_i's as well as of the y_j's. This assumption was removed in the early 1970s by R. Tijdeman [217] (further references, especially to papers by A.A. Smelev and W.D. Brownawell, are given in [230]; see also [35, 235, 236, 242]).

Theorem 13. *Let $x_1, \ldots, x_d$ be $\mathbb{Q}$-linearly independent complex numbers and let $y_1, \ldots, y_\ell$ be $\mathbb{Q}$-linearly independent complex numbers.*
1. If $d\ell \geq 2(d + \ell)$, then at least two of the $d\ell$ numbers

$$e^{x_i y_j} \qquad (1 \leq i \leq d, \ 1 \leq j \leq \ell)$$

are algebraically independent.
2. If $d\ell \geq d + 2\ell$, then at least two of the $d\ell + d$ numbers

$$x_i, \ e^{x_i y_j} \qquad (1 \leq i \leq d, \ 1 \leq j \leq \ell)$$

are algebraically independent.
3. If $d\ell > d + \ell$, then at least two of the $d\ell + d + \ell$ numbers

$$x_i, \ y_j, \ e^{x_i y_j} \qquad (1 \leq i \leq d, \ 1 \leq j \leq \ell)$$

are algebraically independent.
4. If $d = \ell = 2$ and if the two numbers $e^{x_1 y_1}$ and $e^{x_1 y_2}$ are algebraic, then at least two of the six numbers

$$x_1, \ x_2, \ y_1, \ y_2, \ e^{x_2 y_1}, \ e^{x_2 y_2}$$

are algebraically independent.

From the last part of Theorem 13, taking $x_1 = y_1 = i\pi$ and $x_2 = y_2 = 1$, one deduces that at least one of the two following statements is true:
(i) *The number e^{π^2} is transcendental.*
(ii) *The two numbers e and π are algebraically independent.*
One expects that both statements are true.

If it were possible to prove that under the assumptions of Theorem 13, at least two of the eight numbers

$$x_1, \ x_2, \ y_1, \ y_2, \ e^{x_1 y_1}, \ e^{x_1 y_2}, \ e^{x_2 y_1}, \ e^{x_2 y_2}$$

are algebraically independent, one would deduce the algebraic independence of the two numbers π and e^{π} (take $x_1 = 1$, $x_2 = i$, $y_1 = \pi$, $y_2 = i\pi$; see Corollary 48 below).

For results concerning *large transcendence degree*, see Section 5.3 below.

1.2 The Exponential Function

The exponential function

$$\exp : \mathbb{C} \to \mathbb{C}^{\times},$$

$$z \mapsto e^{z},$$

satisfies both a differential equation and an addition formula:

$$\frac{d}{dz} e^{z} = e^{z}, \qquad e^{z_1 + z_2} = e^{z_1} e^{z_2}.$$

It is a homomorphism of the additive group $\mathbb{C}$ of complex numbers onto the multiplicative group $\mathbb{C}^{\times}$ of nonzero complex numbers, with kernel

$$\ker \exp = 2\pi i \mathbb{Z}.$$

Hence it yields an isomorphism between the quotient additive group $\mathbb{C}/2\pi i \mathbb{Z}$ and the multiplicative group $\mathbb{C}^{\times}$.

The group $\mathbb{C}^{\times}$ is the group of complex points of the multiplicative group $\mathbb{G}_m$; $z \mapsto e^{z}$ is the exponential function of the multiplicative group $\mathbb{G}_m$. We shall replace this algebraic group by an elliptic curve. We could replace it also by other commutative algebraic groups. As a first example, the exponential function of the additive group $\mathbb{G}_a$ is

$$\mathbb{C} \to \mathbb{C},$$

$$z \mapsto z.$$

More general examples are commutative linear algebraic groups; over an algebraically closed field, these are nothing else than products of several copies of the additive and multiplicative group. Further examples of algebraic groups are abelian varieties. In full generality, algebraic groups are extensions of abelian varieties by commutative linear algebraic groups. See, for instance, [120, 158, 233].

2 Elliptic Curves and Elliptic Functions

Among many references for this section are the books by S. Lang [127], K. Chandrasekharan [43], and J. Silverman [212, 213]. See also the book by M. Hindry and J. Silverman [99].

2.1 Basic Concepts

An elliptic curve may be defined as

- $y^2 = C(x)$ for a square-free cubic polynomial $C(x)$,
- a connected compact Lie group of dimension 1,
- a complex torus $\mathbb{C}/\Omega$, where Ω is a lattice in $\mathbb{C}$,
- a Riemann surface of genus 1,

– a nonsingular cubic in $\mathbb{P}_2(\mathbb{C})$ (together with a point at infinity),
– an algebraic group of dimension 1, with underlying projective algebraic variety.

We shall use the Weierstrass form

$$E = \{(t : x : y) \; ; \; y^2 t = 4x^3 - g_2 x t^2 - g_3 t^3\} \subset \mathbb{P}_2.$$

Here g_2 and g_3 are complex numbers, with the only assumption $g_2^3 \neq 27 g_3^2$, which means that the discriminant of the polynomial $4X^3 - g_2 X - g_3$ does not vanish.

An analytic parametrization of the complex points $E(\mathbb{C})$ of E is given by means of *the Weierstrass elliptic function* $\wp$, which satisfies the differential equation

$$\wp'^2 = 4\wp^3 - g_2\wp - g_3. \tag{1}$$

It has a double pole at the origin with principal part $1/z^2$ and also satisfies an addition formula

$$\wp(z_1 + z_2) = -\wp(z_1) - \wp(z_2) + \frac{1}{4} \cdot \left(\frac{\wp'(z_1) - \wp'(z_2)}{\wp(z_1) - \wp(z_2)} \right)^2. \tag{2}$$

The exponential map of the Lie group $E(\mathbb{C})$ is

$$\exp_E : \mathbb{C} \to E(\mathbb{C}),$$
$$z \mapsto (1 : \wp(z) : \wp'(z)).$$

The kernel of this map is a *lattice* in $\mathbb{C}$ (that is, a discrete rank-2 subgroup),

$$\Omega = \ker \exp_E = \{\omega \in \mathbb{C} \; ; \; \wp(z + \omega) = \wp(z)\} = \mathbb{Z}\omega_1 + \mathbb{Z}\omega_2.$$

Hence $\exp_E$ induces an isomorphism between the quotient additive group $\mathbb{C}/\Omega$ and $E(\mathbb{C})$ with the law given by (2). The elements of Ω are the *periods* of $\wp$. A pair (ω_1, ω_2) of fundamental periods is given by (cf. [244] Section 20.32, Example 1)

$$\omega_i = 2 \int_{e_i}^{\infty} \frac{dx}{\sqrt{4x^3 - g_2 x - g_3}} \qquad (i = 1, 2),$$

where

$$4x^3 - g_2 x - g_3 = 4(x - e_1)(x - e_2)(x - e_3).$$

Indeed, since $\wp'$ is periodic and odd, it vanishes at $\omega_1/2$, $\omega_2/2$ and $(\omega_1 + \omega_2)/2$; hence the values of $\wp$ at these points are the three distinct complex numbers e_1, e_2, and e_3 (recall that the discriminant of $4x^3 - g_2 x - g_3$ is not 0).

Conversely, given a lattice Ω, there is a unique Weierstrass elliptic function $\wp_\Omega$ whose period lattice is Ω (see Section 2.5). We denote its invariants in the differential equation (1) by $g_2(\Omega)$ and $g_3(\Omega)$.

We shall be interested mainly (but not only) in elliptic curves that are defined over the field of algebraic numbers: they have a Weierstrass equation with algebraic

g_2 and g_3. However, we shall also use the Weierstrass elliptic function associated with the lattice $\lambda\Omega$, where $\lambda \in \mathbb{C}^\times$ may be transcendental; the relations are

$$\wp_{\lambda\Omega}(\lambda z) = \lambda^{-2}\wp_\Omega(z), \qquad g_2(\lambda\Omega) = \lambda^{-4}g_2(\Omega), \qquad g_3(\lambda\Omega) = \lambda^{-6}g_3(\Omega). \quad (3)$$

The lattice $\Omega = \mathbb{Z} + \mathbb{Z}\tau$, where τ is a complex number with positive imaginary part, satisfies

$$g_2(\mathbb{Z} + \mathbb{Z}\tau) = 60G_2(\tau) \quad \text{and} \quad g_3(\mathbb{Z} + \mathbb{Z}\tau) = 140G_3(\tau),$$

where $G_k(\tau)$ (with $k \geq 2$) are the Eisenstein series (see, for instance, [48] Section 3.2, [208] Section 7.2.3, [116] Section 3.2 or [212] Section 6.3 — the normalization in [254] p. 240 is different):

$$G_k(\tau) = \sum_{(m,n)\in\mathbb{Z}^2\setminus\{(0,0)\}} (m + n\tau)^{-2k}. \quad (4)$$

2.2 Morphisms between Elliptic Curves. The Modular Invariant

If Ω and Ω' are two lattices in $\mathbb{C}$ and if $f : \mathbb{C}/\Omega \to \mathbb{C}/\Omega'$ is an analytic homomorphism, then the map $\mathbb{C} \to \mathbb{C}/\Omega \to \mathbb{C}/\Omega'$ factors through a homothecy $\mathbb{C} \to \mathbb{C}$ given by some $\lambda \in \mathbb{C}$ such that $\lambda\Omega \subset \Omega'$:

$$
\begin{array}{ccc}
\mathbb{C} & \xrightarrow{\ \lambda\ } & \mathbb{C} \\
\downarrow & & \downarrow \\
\mathbb{C}/\Omega & \xrightarrow[f]{} & \mathbb{C}/\Omega'
\end{array}
$$

If $f \neq 0$, then $\lambda \in \mathbb{C}^\times$ and f is surjective.

Conversely, if there exists $\lambda \in \mathbb{C}$ such that $\lambda\Omega \subset \Omega'$, then $f_\lambda(x + \Omega) = \lambda x + \Omega'$ defines an analytic surjective homomorphism $f_\lambda : \mathbb{C}/\Omega \to \mathbb{C}/\Omega'$. In this case $\lambda\Omega$ is a subgroup of finite index in Ω'; hence the kernel of f_λ is finite and there exists $\mu \in \mathbb{C}^\times$ with $\mu\Omega' \subset \Omega$: the two elliptic curves $\mathbb{C}/\Omega$ and $\mathbb{C}/\Omega'$ are *isogenous*.

If Ω and Ω^* are two lattices, $\wp$ and $\wp^*$ the associated Weierstrass elliptic functions, and g_2, g_3 the invariants of $\wp$, the following statements are equivalent:

(i) There is a 2×2 matrix with rational coefficients that maps a basis of Ω to a basis of Ω^*.
(ii) There exists $\lambda \in \mathbb{Q}^\times$ such that $\lambda\Omega \subset \Omega^*$.
(iii) There exists $\lambda \in \mathbb{Z} \setminus \{0\}$ such that $\lambda\Omega \subset \Omega^*$.
(iv) The two functions $\wp$ and $\wp^*$ are algebraically dependent over the field $\mathbb{Q}(g_2, g_3)$.
(v) The two functions $\wp$ and $\wp^*$ are algebraically dependent over $\mathbb{C}$.

The map f_λ is an isomorphism if and only if $\lambda\Omega = \Omega'$.
The number

$$j = \frac{1728g_2^3}{g_2^3 - 27g_3^2}$$

is the *modular invariant* of the elliptic curve E. Two elliptic curves over $\mathbb{C}$ are isomorphic if and only if they have the same modular invariant.

Set $\tau = \omega_2/\omega_1$, $q = e^{2\pi i \tau}$ and $J(e^{2\pi i \tau}) = j(\tau)$. Then

$$J(q) = q^{-1} \left(1 + 240 \sum_{m=1}^{\infty} m^3 \frac{q^m}{1-q^m} \right)^3 \prod_{n=1}^{\infty} (1-q^n)^{-24}$$

$$= \frac{1}{q} + 744 + 196884\, q + 21493760\, q^2 + \cdots.$$

See [142] Section 4.12 or [208] Sections 7.3.3 and 7.4.

2.3 Endomorphisms of an Elliptic Curve; Complex Multiplication

Let Ω be a lattice in $\mathbb{C}$. The set of analytic endomorphisms of $\mathbb{C}/\Omega$ is the subring

$$\mathrm{End}(\mathbb{C}/\Omega) = \{ f_\lambda \; ; \; \lambda \in \mathbb{C} \text{ with } \lambda\Omega \subset \Omega \}$$

of $\mathbb{C}$. We also call it the ring of endomorphisms of the associated elliptic curve, or of the corresponding Weierstrass $\wp$ function, and we identify it with the subring

$$\{ \lambda \in \mathbb{C} \; ; \; \lambda\Omega \subset \Omega \}$$

of $\mathbb{C}$. The *field of endomorphisms* is the quotient field $\mathrm{End}(\mathbb{C}/\Omega) \otimes_{\mathbb{Z}} \mathbb{Q}$ of this ring.

If $\lambda \in \mathbb{C}$ satisfies $\lambda\Omega \subset \Omega$, then λ is either a rational integer or an algebraic integer in an imaginary quadratic field. For such a λ, $\wp_\Omega(\lambda z)$ is a rational function of $\wp_\Omega(z)$; the degree of the numerator is λ^2 if $\lambda \in \mathbb{Z}$ and $N(\lambda)$ otherwise (here, N is the norm of the imaginary quadratic field); the degree of the denominator is $\lambda^2 - 1$ if $\lambda \in \mathbb{Z}$ and $N(\lambda) - 1$ otherwise.

Let E be the elliptic curve attached to the Weierstrass $\wp$ function. The ring of endomorphisms $\mathrm{End}(E)$ of E is either $\mathbb{Z}$ or an order in an imaginary quadratic field k. The latter case arises if and only if the quotient $\tau = \omega_2/\omega_1$ of a pair of fundamental periods is a quadratic number; in this case the field of endomorphisms of E is $k = \mathbb{Q}(\tau)$ and the curve E has *complex multiplication*; this is the so-called *CM case*. This means also that the two functions $\wp(z)$ and $\wp(\tau z)$ are algebraically dependent. In this case, the value $j(\tau)$ of the modular invariant j is an algebraic integer whose degree is the class number of the quadratic field $k = \mathbb{Q}(\tau)$.

Remark 14. From Theorem 7 one deduces the transcendence of the number

$$e^{\pi\sqrt{163}} = 262\,537\,412\,640\,768\,743.999\,999\,999\,999\,250\,072\,59\ldots.$$

If we set

$$\tau = \frac{1 + i\sqrt{163}}{2}, \quad q = e^{2\pi i \tau} = -e^{-\pi\sqrt{163}},$$

then the class number of the imaginary quadratic field $\mathbb{Q}(\tau)$ is 1, we have $j(\tau) = -(640\,320)^3$, and

$$\left| j(\tau) - \frac{1}{q} - 744 \right| < 10^{-12}.$$

Also ([57] Section 2.4)

$$\left(e^{\pi\sqrt{163}} - 744 \right)^{1/3} = 640\,319.999\,999\,999\,999\,999\,999\,999\,999\,390\,31\ldots.$$

Let $\wp$ be a Weierstrass elliptic function with field of endomorphisms k. Hence $k = \mathbb{Q}$ if the associated elliptic curve has no complex multiplication, while in the other case k is an imaginary quadratic field, namely $k = \mathbb{Q}(\tau)$, where τ is the quotient of two linearly independent periods of $\wp$. Let $u_1, \ldots, u_d$ be nonzero complex numbers. Then the functions $\wp(u_1 z), \ldots, \wp(u_d z)$ are algebraically independent (over $\mathbb{C}$ or over $\mathbb{Q}(g_2, g_3)$; this is equivalent) if and only if the numbers $u_1, \ldots, u_d$ are linearly independent over k. This generalizes the fact that $\wp(z)$ and $\wp(\tau z)$ are algebraically dependent if and only if the elliptic curve has complex multiplication. Much more general and deeper results of algebraic independence of functions (exponential and elliptic functions, zeta functions, ...) were proved by W.D. Brownawell and K.K. Kubota [37].

If $\wp$ is a Weierstrass elliptic function with algebraic invariants g_2 and g_3, if E is the associated elliptic curve, and if k denotes its field of endomorphisms, then the set

$$\mathcal{L}_E = \Omega \cup \left\{ u \in \mathbb{C} \setminus \Omega \; ; \; \wp(u) \in \overline{\mathbb{Q}} \right\}$$

is a k-vector subspace of $\mathbb{C}$: this is the set of *elliptic logarithms of algebraic points on E*. It plays a role with respect to E similar to the role of $\mathcal{L}$ for the multiplicative group $\mathbb{G}_m$.

Let $k = \mathbb{Q}(\sqrt{-d})$ be an imaginary quadratic field with class number $h(-d) = h$. There are h nonisomorphic elliptic curves $E_1, \ldots, E_h$ with ring of endomorphisms the ring of integers of k. The numbers $j(E_i)$ are conjugate algebraic integers of degree h; each of them generates the Hilbert class field H of k (maximal unramified abelian extension of k). The Galois group of H/k is isomorphic to the ideal class group of k.

Since the group of roots of units of an imaginary quadratic field is $\{-1, +1\}$ except for $\mathbb{Q}(i)$ and $\mathbb{Q}(\varrho)$, where $\varrho = e^{2\pi i/3}$, it follows that there are exactly two elliptic curves over $\mathbb{Q}$ (up to isomorphism) having an automorphism group bigger than $\{-1, +1\}$. They correspond to Weierstrass elliptic functions $\wp$ for which there exists a complex number $\lambda \neq \pm 1$ with $\lambda^2 \wp(\lambda z) = \wp(z)$.

The first one has $g_3 = 0$ and $j = 1728$. An explicit value for a pair of fundamental periods of the elliptic curve

$$y^2 t = 4x^3 - 4xt^2$$

follows from computations by Legendre using Gauss's lemniscate function ([244] Section 22.8) and yields (see [4], as well as Appendix 1 of [241])

$$\omega_1 = \int_1^\infty \frac{dx}{\sqrt{x^3 - x}} = \frac{1}{2} B(1/4, 1/2) = \frac{\Gamma(1/4)^2}{2^{3/2}\pi^{1/2}} \quad \text{and} \quad \omega_2 = i\omega_1. \quad (5)$$

The lattice $\mathbb{Z}[i]$ has $g_2 = 4\omega_1^4$. Thus

$$\sum_{(m,n)\in\mathbb{Z}^2\setminus\{(0,0)\}} (m+ni)^{-4} = \frac{\Gamma(1/4)^8}{2^6 \cdot 3 \cdot 5 \cdot \pi^2}.$$

The second one has $g_2 = 0$ and $j = 0$. Again from computations by Legendre ([244] Section 22.81 II) one deduces that a pair of fundamental periods of the elliptic curve

$$y^2 t = 4x^3 - 4t^3$$

is (see once more [4] and Appendix 1 of [241])

$$\omega_1 = \int_1^\infty \frac{dx}{\sqrt{x^3 - 1}} = \frac{1}{3} B(1/6, 1/2) = \frac{\Gamma(1/3)^3}{2^{4/3}\pi} \quad \text{and} \quad \omega_2 = \varrho\omega_1. \quad (6)$$

The lattice $\mathbb{Z}[\varrho]$ has $g_3 = 4\omega_1^6$. Thus

$$\sum_{(m,n)\in\mathbb{Z}^2\setminus\{(0,0)\}} (m+n\varrho)^{-6} = \frac{\Gamma(1/3)^{18}}{2^8 \cdot 5 \cdot 7 \cdot \pi^6}.$$

These two examples involve special values of Euler's gamma function

$$\Gamma(z) = \int_0^\infty e^{-t} t^z \cdot \frac{dt}{t} = e^{-\gamma z} z^{-1} \prod_{n=1}^\infty \left(1 + \frac{z}{n}\right)^{-1} e^{z/n}, \quad (7)$$

where

$$\gamma = \lim_{n\to\infty} \left(\sum_{k=1}^n \frac{1}{k} - \log n\right) = 0.577\,215\,664\,901\,532\,860\,606\,512\,09\ldots$$

is Euler's constant (Section 12.1 in [244]), while Euler's beta function is

$$B(a,b) = \frac{\Gamma(a)\Gamma(b)}{\Gamma(a+b)} = \int_0^1 x^{a-1}(1-x)^{b-1}dx.$$

More generally, the formula of Chowla and Selberg (1966) [44] (see also [9, 95, 96, 115, 117, 234] for related results) expresses periods of elliptic curves with complex multiplication as products of gamma values: *if k is an imaginary quadratic field and $\mathcal{O}$ an order in k, if E is an elliptic curve with complex multiplication by $\mathcal{O}$, then the corresponding lattice Ω determines a vector space $\Omega \otimes_\mathbb{Z} \mathbb{Q}$ that is invariant under the action of k and thus has the form $k \cdot \omega$ for some $\omega \in \mathbb{C}^\times$ defined up to elements in $k^\times$. In particular, if $\mathcal{O}$ is the ring of integers $\mathbb{Z}_k$ of k, then*

$$\omega = \alpha\sqrt{\pi} \prod_{\substack{0<a<d \\ (a,d)=1}} \Gamma(a/d)^{w\epsilon(a)/4h},$$

where α is a nonzero algebraic number, w is the number of roots of unity in k, h is the class number of k, and ϵ is the Dirichlet character modulo the discriminant d of k.

2.4 Standard Relations among Gamma Values

Euler's gamma function satisfies the following relations ([244] Chapter XII):
(Translation)

$$\Gamma(z + 1) = z\Gamma(z);$$

(Reflection)

$$\Gamma(z)\Gamma(1 - z) = \frac{\pi}{\sin(\pi z)};$$

(Multiplication) For any positive integer n,

$$\prod_{k=0}^{n-1} \Gamma\left(z + \frac{k}{n}\right) = (2\pi)^{(n-1)/2} n^{-nz+(1/2)} \Gamma(nz).$$

D. Rohrlich conjectured that any multiplicative relation among gamma values is a consequence of these standard relations, while S. Lang was more optimistic (see [125], [128] I Chapter 2 Appendix p. 66 and [9] Chapter 24):

Conjecture 15 (D. Rohrlich). Any multiplicative relation

$$\pi^{b/2} \prod_{a \in \mathbb{Q}} \Gamma(a)^{m_a} \in \overline{\mathbb{Q}}$$

with b and m_a in $\mathbb{Z}$ is a consequence of the standard relations.

Conjecture 16 (S. Lang). Any algebraic dependence relation with algebraic coefficients among the numbers $(2\pi)^{-1/2}\Gamma(a)$ with $a \in \mathbb{Q}$ is in the ideal generated by the standard relations.

2.5 Quasiperiods of Elliptic Curves and Elliptic Integrals of the Second Kind

Let $\Omega = \mathbb{Z}\omega_1 + \mathbb{Z}\omega_2$ be a lattice in $\mathbb{C}$. The *Weierstrass canonical product* attached to this lattice is the entire function σ_Ω defined by ([244] Section 20.42)

$$\sigma_\Omega(z) = z \prod_{\omega \in \Omega \setminus \{0\}} \left(1 - \frac{z}{\omega}\right) e^{\frac{z}{\omega} + \frac{z^2}{2\omega^2}}.$$

It has a simple zero at any point of Ω.

Hence the Weierstrass sigma function plays, for the lattice Ω, the role that is played by the function

$$z \prod_{n \geq 1} \left(1 - \frac{z}{n}\right) e^{z/n} = -e^{\gamma z} \Gamma(-z)^{-1}$$

for the set of positive integers $\mathbb{N} \setminus \{0\} = \{1, 2, \ldots\}$ (see the infinite product (7) for Euler's gamma function), and also by the function

$$\pi^{-1} \sin(\pi z) = z \prod_{n \in \mathbb{Z} \setminus \{0\}} \left(1 - \frac{z}{n}\right) e^{z/n}$$

for the set $\mathbb{Z}$ of rational integers ([43] Section 4.2).

The Weierstrass sigma function σ associated with a lattice in $\mathbb{C}$ is an entire function of *order* 2:

$$\limsup_{r \to \infty} \frac{1}{\log r} \cdot \log \log \sup_{|z|=r} |\sigma(z)| = 2;$$

the product $\sigma^2 \wp$ is also an entire function of order 2 (this can be checked using infinite products, but it is easier to use the quasiperiodicity of σ, see formula (8) below).

The logarithmic derivative of the sigma function is *the Weierstrass zeta function* $\zeta = \sigma'/\sigma$ whose Laurent expansion at the origin is ([127] Section 18.3, [208] Section 7.2.3 and [212] Section 6.3, Theorem 3.5)

$$\zeta(z) = \frac{1}{z} - \sum_{k \geq 2} s_k z^{2k-1},$$

where for $k \in \mathbb{Z}, k \geq 2$,

$$s_k = s_k(\Omega) = \sum_{\substack{\omega \in \Omega \\ \omega \neq 0}} \omega^{-2k} = \omega_1^{-2k} G_k(\tau)$$

(recall (4); also $\tau = \omega_2/\omega_1$).

The derivative of ζ is $-\wp$. From

$$\wp'' = 6\wp^2 - (g_2/2)$$

one deduces that $s_k(\Omega)$ is a homogeneous polynomial in $\mathbb{Q}[g_2, g_3]$ of weight $2k$ for the graduation of $\mathbb{Q}[g_2, g_3]$ determined by assigning to g_2 the degree 4 and to g_3 the degree 6.

As a side remark, we notice that for any $u \in \mathbb{C} \setminus \Omega$ we have

$$\mathbb{Q}(g_2, g_3) \subset \mathbb{Q}\big(\wp(u), \wp'(u), \wp''(u)\big).$$

Since its derivative is periodic, the function ζ is *quasiperiodic*: for each $\omega \in \Omega$ there is a complex number $\eta = \eta(\omega)$ such that

$$\zeta(z + \omega) = \zeta(z) + \eta.$$

These numbers η are the *quasiperiods* of the elliptic curve. If (ω_1, ω_2) is a pair of fundamental periods and if we set $\eta_1 = \eta(\omega_1)$ and $\eta_2 = \eta(\omega_2)$, then, for $(a, b) \in \mathbb{Z}^2$,

$$\eta(a\omega_1 + b\omega_2) = a\eta_1 + b\eta_2.$$

Returning to the sigma function, one deduces that

$$\sigma(z + \omega_i) = -\sigma(z) \exp(\eta_i(z + (\omega_i/2))) \qquad (i = 1, 2). \tag{8}$$

The zeta function also satisfies an addition formula:

$$\zeta(z_1 + z_2) = \zeta(z_1) + \zeta(z_2) + \frac{1}{2} \cdot \frac{\wp'(z_1) - \wp'(z_2)}{\wp(z_1) - \wp(z_2)}.$$

The Legendre relation relating the periods and the quasiperiods

$$\omega_2 \eta_1 - \omega_1 \eta_2 = 2\pi i,$$

when ω_2/ω_1 has positive imaginary part, can be obtained by integrating $\zeta(z)$ along the boundary of a fundamental parallelogram ([43] Section 4.2, [124] Section 1.6, [244] Section 20.411).

In the case of complex multiplication, if τ is the quotient of a pair of fundamental periods of $\wp$, then the function $\zeta(\tau z)$ is algebraic over the field $\mathbb{Q}(g_2, g_3, z, \wp(z), \zeta(z))$.

Examples ([4, 241]). For the curve $y^2 t = 4x^3 - 4xt^2$, the quasiperiods attached to the above-mentioned pair of fundamental periods (5) are

$$\eta_1 = \frac{\pi}{\omega_1} = \frac{(2\pi)^{3/2}}{\Gamma(1/4)^2}, \qquad \eta_2 = -i\eta_1; \tag{9}$$

it follows that the fields $\mathbb{Q}(\omega_1, \omega_2, \eta_1, \eta_2)$ and $\mathbb{Q}(\pi, \Gamma(1/4))$ have the same algebraic closure over $\mathbb{Q}$, hence the same transcendence degree. For the curve $y^2 t = 4x^3 - 4t^3$ with periods (6), they are

$$\eta_1 = \frac{2\pi}{\sqrt{3}\omega_1} = \frac{2^{7/3}\pi^2}{3^{1/2}\Gamma(1/3)^3}, \qquad \eta_2 = \varrho^2 \eta_1. \tag{10}$$

In this case the fields $\mathbb{Q}(\omega_1, \omega_2, \eta_1, \eta_2)$ and $\mathbb{Q}(\pi, \Gamma(1/3))$ have the same algebraic closure over $\mathbb{Q}$, hence the same transcendence degree.

2.6 Elliptic Integrals

Let

$$\mathcal{E} = \{(t : x : y) \in \mathbb{P}_2; \, y^2 t = 4x^3 - g_2 x t^2 - g_3 t^3\}$$

be an elliptic curve. The field of rational (meromorphic) functions on $\mathcal{E}$ over $\mathbb{C}$ is $\mathbb{C}(\mathcal{E}) = \mathbb{C}(\wp, \wp') = \mathbb{C}(x, y)$, where x and y are related by the cubic equation $y^2 = 4x^3 - g_2 x - g_3$. Under the isomorphism $\mathbb{C}/\Omega \to \mathcal{E}(\mathbb{C})$ given by $(1 : \wp : \wp')$, the differential form dz is mapped to dx/y. The holomorphic differential forms on $\mathbb{C}/\Omega$ are $\lambda \, dz$ with $\lambda \in \mathbb{C}$.

The differential form $d\zeta = \zeta'/\zeta$ is mapped to $-x\,dx/y$. The differential forms of the second kind on $\mathcal{E}(\mathbb{C})$ are $a\,dz + b\,d\zeta + d\chi$, where a and b are complex numbers and $\chi \in \mathbb{C}(x, y)$ is a meromorphic function on $\mathcal{E}$.

Assume that the elliptic curve $\mathcal{E}$ is defined over $\overline{\mathbb{Q}}$: the invariants g_2 and g_3 are algebraic. We shall be interested in differential forms defined over $\overline{\mathbb{Q}}$. Those of the second kind are $a\,dz + b\,d\zeta + d\chi$, where a and b are algebraic numbers and $\chi \in \overline{\mathbb{Q}}(x, y)$.

An elliptic integral (see [244] Section 22.7; see also [43] Section 1.4 and [212] Section 6.1) is an integral

$$\int R(x, y)dx,$$

where R is a rational function of x and y, while y^2 is a polynomial in x of degree 3 or 4 without multiple roots, with the proviso that the integral cannot be integrated by means of elementary functions. One may transform this integral as follows: one reduces it to an integral of $dx/\sqrt{P(x)}$, where P is a polynomial of third or fourth degree; in case P has degree 4, one replaces it with a degree-3 polynomial by sending one root to infinity; finally, one reduces it to a Weierstrass equation by means of a birational transformation. The value of the integral is not modified.

For transcendence purposes, if the initial differential form is defined over $\overline{\mathbb{Q}}$, then all these transformations involve only algebraic numbers.

We refer to Section 22.7 of [244] for the definition of elliptic integrals of the first, second, and third kinds.

3 Transcendence Results of Numbers Related to Elliptic Functions

3.1 Elliptic Analogue of Lindemann's Theorem on the Transcendence of π and the Hermite–Lindemann Theorem on the Transcendence of $\log \alpha$

The first transcendence result on periods of elliptic functions was proved by C.L. Siegel [210] as early as 1932.

Theorem 17 (Siegel, 1932). *Let $\wp$ be a Weierstrass elliptic function with period lattice $\mathbb{Z}\omega_1 + \mathbb{Z}\omega_2$. Assume that the invariants g_2 and g_3 of $\wp$ are algebraic. Then at least one of the two numbers ω_1, ω_2 is transcendental.*

One main feature of Siegel's proof is that he used Dirichlet's box principle (the so-called Thue–Siegel lemma, which is included in his 1929 paper) to construct an auxiliary function. This idea turned out to be of fundamental importance for the solution of Hilbert's seventh problem by Gel'fond and Schneider two years later.

In the case of complex multiplication, it follows from Theorem 17 that *any nonzero period of $\wp$ is transcendental.*

From formulas (5) and (6) it follows as a consequence of Siegel's 1932 result [210] that both numbers $\Gamma(1/4)^4/\pi$ and $\Gamma(1/3)^3/\pi$ are transcendental.

Other consequences of Siegel's result concern the transcendence of the length of an arc of an ellipse [207, 211]:

$$2 \int_{-b}^{b} \sqrt{1 + \frac{a^2 x^2}{b^4 - b^2 x^2}} \, dx$$

for algebraic a and b, as well as the transcendence of an arc of the lemniscate $(x^2 + y^2)^2 = 2a^2(x^2 - y^2)$ with a algebraic.

A further example of application of Siegel's theorem [211] is the transcendence of values of hypergeometric series related to elliptic integrals

$$K(z) = \int_{0}^{1} \frac{dx}{\sqrt{(1 - x^2)(1 - z^2 x^2)}} = \frac{\pi}{2} \cdot {}_2F_1(1/2, \ 1/2 \ ; \ 1 \mid z^2),$$

where ${}_2F_1$ denotes the Gauss hypergeometric series

$$ {}_2F_1\left(a, \ b \ ; \ c \mid z\right) = \sum_{n=0}^{\infty} \frac{(a)_n (b)_n}{(c)_n} \cdot \frac{z^n}{n!}$$

with $(a)_n = a(a+1) \cdots (a + n - 1)$.

Further results on this topic were obtained by Th. Schneider [203] in 1934 and in joint work by K. Mahler and J. Popken [190] in 1935 using Siegel's method. These results were superseded by Th. Schneider's fundamental memoir [204] in 1936 in which he proved a number of definitive results on the subject, including the following:

Theorem 18 (Schneider, 1936). *Assume that the invariants g_2 and g_3 of $\wp$ are algebraic. Then for any nonzero period ω of $\wp$, the numbers ω and $\eta(\omega)$ are transcendental.*

It follows from Theorem 18 that any nonzero period of an elliptic integral of the first or second kind is transcendental:

Corollary 19. *Let $\mathcal{E}$ be an elliptic curve over $\overline{\mathbb{Q}}$, p_1 and p_2 two algebraic points on $\mathcal{E}(\overline{\mathbb{Q}})$, w a differential form of the first or second kind on $\mathcal{E}$ that is defined over $\overline{\mathbb{Q}}$, holomorphic at p_1 and p_2, and is not the differential of a rational function. Let γ be a path on $\mathcal{E}$ from p_1 to p_2. In case $p_1 = p_2$ one assumes that γ is not homologous to 0. Then the number*

$$\int_{\gamma} w$$

is transcendental.

Examples. Using Corollary 19 and formulas (9) and (10), one deduces that the numbers

$$\Gamma(1/4)^4 / \pi^3 \quad \text{and} \quad \Gamma(1/3)^3 / \pi^2$$

are transcendental.

The main results of Schneider's 1936 paper [204] are as follows (see also [207]):

Theorem 20 (Schneider, 1936).
1. *Let $\wp$ be a Weierstrass elliptic function with algebraic invariants g_2, g_3. Let β be a nonzero algebraic number. Then β is not a pole of $\wp$ and $\wp(\beta)$ is transcendental. More generally, if a and b are two algebraic numbers with $(a, b) \neq (0, 0)$, then for any $u \in \mathbb{C} \setminus \Omega$ at least one of the two numbers $\wp(u)$, $au + b\zeta(u)$ is transcendental.*
2. *Let $\wp$ and $\wp^*$ be two algebraically independent elliptic functions with algebraic invariants g_2, g_3, g_2^*, g_3^*. If $t \in \mathbb{C}$ is not a pole of $\wp$ or of $\wp^*$, then at least one of the two numbers $\wp(t)$ and $\wp^*(t)$ is transcendental.*
3. *Let $\wp$ be a Weierstrass elliptic function with algebraic invariants g_2, g_3. Then for any $t \in \mathbb{C} \setminus \Omega$, at least one of the two numbers $\wp(t)$, e^t is transcendental.*

It follows from Theorem 20.2 that the quotient of an elliptic integral of the first kind (between algebraic points) by a nonzero period is either in the field of endomorphisms (hence a rational number, or a quadratic number in the field of complex multiplication), or a transcendental number.

Here is another important consequence of Theorem 20.2.

Corollary 21 (Schneider, 1936). *Let $\tau \in \mathcal{H}$ be a complex number in the upper half-plane $\Im m(\tau) > 0$ such that $j(\tau)$ is algebraic. Then τ is algebraic if and only if τ is imaginary quadratic.*

In this connection we quote Schneider's second problem in [207], which is still open (see Wakabayashi's papers [226, 227, 228]):

Open Problem. Prove Corollary 21 without using elliptic functions.

Sketch of proof of Corollary 21 as a consequence of part 2 of Theorem 20. Assume that both $\tau \subset \mathcal{H}$ and $j(\tau)$ are algebraic. There exists an elliptic function with algebraic invariants g_2, g_3 and periods ω_1, ω_2 such that

$$\tau = \frac{\omega_2}{\omega_1} \quad \text{and} \quad j(\tau) = \frac{1728 g_2^3}{g_2^3 - 27 g_3^2}.$$

Set $\wp^*(z) = \tau^2 \wp(\tau z)$. Then $\wp^*$ is a Weierstrass function with algebraic invariants g_2^*, g_3^*. For $u = \omega_1/2$ the two numbers $\wp(u)$ and $\wp^*(u)$ are algebraic. Hence the two functions $\wp(z)$ and $\wp^*(z)$ are algebraically dependent. It follows that the corresponding elliptic curve has nontrivial endomorphisms; therefore τ is quadratic. $\qquad\square$

A quantitative refinement of Schneider's theorem on the transcendence of $j(\tau)$ given by A. Faisant and G. Philibert in 1984 [74] became useful 10 years later in connection with Nesterenko's result (see Section 5). See also [75].

We will not review the results related to abelian integrals, but only quote the first result on this topic, which involves the Jacobian of a Fermat curve: in 1941 Schneider [205] proved that *for a and b in $\mathbb{Q}$ with a, b and $a + b$ not in $\mathbb{Z}$, the number*

$$B(a, b) = \frac{\Gamma(a)\Gamma(b)}{\Gamma(a + b)}$$

is transcendental. We notice that in his 1932 paper [210], C.L. Siegel had already announced partial results on the values of the Euler gamma function (see also [19]).

Schneider's above-mentioned results deal with elliptic (and abelian) integrals of the first or second kind. His method can be extended to deal with elliptic (and abelian) integrals of the third kind (this is Schneider's third problem in [207]).

As pointed out by J.-P. Serre in 1979 [233], it follows from the quasiperiodicity of the Weierstrass sigma function (8) that the function

$$F_u(z) = \frac{\sigma(z + u)}{\sigma(z)\sigma(u)} e^{-z\zeta(u)}$$

satisfies

$$F_u(z + \omega_i) = F_u(z) e^{\eta_i u - \omega_i \zeta(u)}.$$

Theorem 22. *Let u_1 and u_2 be two nonzero complex numbers. Assume that g_2, g_3, $\wp(u_1)$, $\wp(u_2)$, β are algebraic and $\mathbb{Z}u_1 \cap \Omega = \{0\}$. Then the number*

$$\frac{\sigma(u_1 + u_2)}{\sigma(u_1)\sigma(u_2)} e^{\left(\beta - \zeta(u_1)\right)u_2}$$

is transcendental.

From the next corollary, one can deduce that nonzero periods of elliptic integrals of the third kind are transcendental (see [232]).

Corollary 23. *For any nonzero period ω and for any $u \in \mathbb{C} \setminus \Omega$, the number $e^{\omega\zeta(u) - \eta u + \beta\omega}$ is transcendental.*

Further results on elliptic integrals are due to M. Laurent [132]. See also his papers [134, 135, 136, 137].

Ya. M. Kholyavka wrote several papers devoted to the approximation of transcendental numbers related to elliptic functions [106, 107, 108, 109, 110, 111, 112, 113, 114].

Quantitative estimates (measures of transcendence) related to the results of this section were derived by N.I. Fel'dman [76, 77, 78, 79, 80]. See also the papers by S. Lang [119], N.D. Nagaev [165], N. Hirata [101], E. Reyssat [195, 196, 198, 199], M. Laurent [133], R. Tubbs [219], G. Diaz [64], N. Saradha [202], P. Grinspan [94].

3.2 Elliptic Analogues of the Six Exponentials Theorem

Elliptic analogs of the six exponentials theorem (Theorem 10) were considered by S. Lang [120] and K. Ramachandra [191] in the 1960s.

Let d_1, d_2 be nonnegative integers and m a positive integer, let $x_1, \ldots, x_{d_1}$ be complex numbers that are linearly independent over $\mathbb{Q}$, let $y_1, \ldots, y_m$ be complex numbers that are linearly independent over $\mathbb{Q}$, and let $u_1, \ldots, u_{d_2}$ be nonzero complex numbers. We consider Weierstrass elliptic functions $\wp_1, \ldots, \wp_{d_2}$ and we denote by K_0 the field generated over $\mathbb{Q}$ by their invariants $g_{2,k}$ and $g_{3,k}$ ($1 \leq k \leq d_2$). We assume that the d_2 functions $\wp_1(u_1 z), \ldots, \wp_{d_2}(u_{d_2} z)$ are algebraically independent. We denote by K_1 the field generated over K_0 by the numbers $\exp(x_i y_j)$ ($1 \leq i \leq d_1, 1 \leq j \leq m$) together with the numbers $\wp_k(u_k y_j)$ ($1 \leq k \leq d_2$, $1 \leq j \leq m$). Next, define

$$K_2 = K_1(y_1, \ldots, y_m), \qquad K_3 = K_1(x_1, \ldots, x_{d_1}, u_1, \ldots, u_{d_2}),$$

and let K_4 be the compositum of K_2 and K_3:

$$K_4 = K_1(y_1, \ldots, y_m, x_1, \ldots, x_{d_1}, u_1, \ldots, u_{d_2}).$$

The theorems of Hermite–Lindemann (Theorem 3), Gel'fond–Schneider (Theorem 7), the six exponentials theorem, and their elliptic analogues due to Schneider, Lang, and Ramachandra can be stated as follows.

Any one of the four assumptions below will imply $d_1 + d_2 > 0$; the case in which d_1 (respectively d_2) vanishes means that one considers only elliptic (respectively exponential) functions.

Theorem 24.
1. Assume $(d_1 + d_2)m > m + d_1 + 2d_2$. Then the field K_1 has transcendence degree ≥ 1 over $\mathbb{Q}$.
2. Assume either $d_1 \geq 1$ and $m \geq 2$, or $d_2 \geq 1$ and $m \geq 3$. Then K_2 has transcendence degree ≥ 1 over $\mathbb{Q}$.
3. Assume $d_1 + d_2 \geq 2$. Then K_3 has transcendence degree ≥ 1 over $\mathbb{Q}$.
4. Assume $d_1 + d_2 \geq 1$. Then K_4 has transcendence degree ≥ 1 over $\mathbb{Q}$.

Parts 3 and 4 of Theorem 24 are consequences of the Schneider–Lang criterion [120], which deals with meromorphic functions satisfying differential equations, while parts 1 and 2 follow from a criterion that involves no differential equations. Such criteria were given by Schneider [206, 207], Lang [120], and Ramachandra [191] (see also [228] and [229]).

Theorem 24 also includes Theorem 20 apart from the case $b \neq 0$ in part 1 of that statement. However, there are extensions of Theorem 24 that include results on Weierstrass zeta functions (and also on Weierstrass sigma functions in connection with elliptic integrals of the third kind). See [132, 134, 135, 136, 137, 199, 232, 233].

Here is a corollary of part 1 of Theorem 24 (take $d_1 = 0$, $d_2 = 3$, $\wp_1 = \wp_2 = \wp_3 = \wp$, $m = 4$, $y_1 = 1$, $y_2 \in \operatorname{End}(E) \setminus \mathbb{Q}$, $y_3 = v_1/u_1$, $y_4 = y_2 y_3$; there is an alternative proof with $d_2 = 2$ and $m = 6$).

Corollary 25. *Let E be an elliptic curve with algebraic invariants g_2, g_3. Assume that E has complex multiplication. Let*

$$M = \begin{pmatrix} u_1 & u_2 & u_3 \\ v_1 & v_2 & v_3 \end{pmatrix}$$

be a 2×3 *matrix whose entries are elliptic logarithms of algebraic numbers, i.e.,* u_i *and* v_i *(*$i = 1, 2, 3$*) are in* $\mathcal{L}_E$*. Assume that the three columns are linearly independent over* $\mathrm{End}(E)$ *and the two rows are also linearly independent over* $\mathrm{End}(E)$*. Then the matrix M has rank 2.*

In the non-CM case, one deduces from Theorem 24 a similar (but weaker) statement according to which such matrices $\left(u_{ij}\right)$ (where $\wp\left(u_{ij}\right)$ are algebraic numbers) have rank ≥ 2 if they have size 2×5 (taking $d_1 = 0$, $d_2 = 2$, and $m = 5$) or 3×4 (taking $d_1 = 0$ and either $d_2 = 3$, $m = 4$ or $d_2 = 4$ and $m = 3$) instead of 2×3.

Lower bounds better than 2 for the rank of matrices of larger sizes are known, but we will not discuss this question here. We just mention the fact that higher-dimensional considerations are relevant to a problem discussed by B. Mazur on the density of rational points on varieties [240].

4 Linear Independence of Numbers Related to Elliptic Functions

From Schneider's theorem (Theorem 20) part 1, one deduces the linear independence over the field of algebraic numbers of the three numbers 1, ω, and η, when ω is a nonzero period of a Weierstrass elliptic function (with algebraic invariants g_2 and g_3) and $\eta = \eta(\omega)$ is the associated quasiperiod of the corresponding Weierstrass zeta function. However, the Gel'fond–Schneider method in one variable alone does not yield strong results of linear independence. Baker's method is better suited for this purpose.

4.1 Linear Independence of Periods and Quasiperiods

Baker's method of proof for his theorem (Theorem 9) on linear independence of logarithms of algebraic numbers was used as early as 1969 and 1970 by A. Baker himself [12, 10] when he proved the transcendence of linear combinations with algebraic coefficients of the numbers ω_1, ω_2, η_1, and η_2 associated with an elliptic curve having algebraic invariants g_2 and g_3. His method is effective: it provides quantitative Diophantine estimates [11].

In 1971, J. Coates [52] proved the transcendence of linear combinations with algebraic coefficients of ω_1, ω_2, η_1, η_2, and $2\pi i$. Moreover, he proved in [51, 53, 54, 55] that in the non-CM case, the three numbers ω_1, ω_2, and $2\pi i$ are $\overline{\mathbb{Q}}$-linearly independent. Further results including usual logarithms of algebraic numbers are due to T. Harase in 1974 and 1976 [97, 98].

The final result on the question of linear dependence of periods and quasiperiods for a single elliptic function was given by D.W. Masser in 1975 [143, 144].

Theorem 26 (Masser, 1975). *Let* $\wp$ *be a Weierstrass elliptic function with algebraic invariants* g_2 *and* g_3*, denote by* ζ *the corresponding Weierstrass zeta function, let* ω_1*,* ω_2 *be a basis of the period lattice of* $\wp$*, and let* η_1*,* η_2 *be the associated quasiperiods of* ζ*. Then the six numbers* 1*,* ω_1*,* ω_2*,* η_1*,* η_2*, and* $2\pi i$ *span a* $\overline{\mathbb{Q}}$*-vector space of dimension 6 in the non-CM case, 4 in the CM case:*

$$\dim_{\overline{\mathbb{Q}}}\{1, \omega_1, \omega_2, \eta_1, \eta_2, 2\pi i\} = 2 + 2 \dim_{\overline{\mathbb{Q}}}\{\omega_1, \omega_2\}.$$

The fact that the dimension is 4 in the CM case means that there are two independent linear relations among these six numbers. One of them is $\omega_2 = \tau\omega_1$ with $\tau \in \overline{\mathbb{Q}}$. The second one (see [144]; see also [37]) can be written

$$C^2\tau\eta_2 - AC\eta_1 + \gamma\omega_1 = 0,$$

where $A + BX + CX^2$ is the minimal polynomial of τ over $\mathbb{Z}$ and γ is an element in $\mathbb{Q}(g_2, g_3, \tau)$.

In [144], D.W. Masser also produced quantitative estimates (measures of linear independence). In 1976, R. Franklin and D.W. Masser [85, 151] obtained an extension involving a logarithm of an algebraic number.

Further results can be found in papers by P. Bundschuh [40], S. Lang's surveys [121, 122], D.W. Masser [152, 154], M. Anderson [5], and in the joint paper [6] by M. Anderson and D.W. Masser.

4.2 Elliptic Analogue of Baker's Theorem

The elliptic analogue of Baker's theorem on linear independence of logarithms was proved by D.W. Masser in 1974 [143, 144] in the CM case.

His proof also yields quantitative estimates (measures of linear independence of elliptic logarithms of algebraic points on an elliptic curve). Such estimates have a number of applications: this was shown by A.O. Gel'fond for usual logarithms of algebraic numbers [89], and further consequences of such lower bounds in the case of elliptic curves for solving Diophantine equations (integer points on elliptic curves) were derived by S. Lang [126].

Lower bounds for linear combinations of elliptic logarithms in the CM case were obtained by several mathematicians including J. Coates [52], D.W. Masser [145, 149, 150], J. Coates and S. Lang [56], M. Anderson [5]. The work of Yu Kunrui [253] yields similar estimates, but his method is not that of Baker–Masser: instead of using a generalization of Gel'fand's solution to Hilbert's seventh problem, Yu Kunrui uses a generalization in several variables of Schneider's solution to the same problem. Again, this method is restricted to the CM case.

The question of linear independence of elliptic logarithms in the non-CM case was settled only in 1980 by D. Bertrand and D.W. Masser [30, 31]. They found a new proof of Baker's theorem using functions of several variables, and they were able to extend this argument to the situation of elliptic functions, either with or without complex multiplication. The criterion they use is the one that Schneider established in 1949 [205] for his proof of the transcendence of beta values. This criterion (revisited by S. Lang in [120]) deals with Cartesian products. From the several variables point of view, this is a rather degenerate situation; much deeper results are available, including Bombieri's solution in 1970 of Nagata's conjecture [120, 233], which involves Hörmander L^2-estimates for analytic functions of several variables. However Bombieri's theorem does not seem to yield new transcendence results, so far.

So far, these deeper results do not give further transcendence results in our context.

Theorem 27 (D.W. Masser 1974 for the CM case, D. Bertrand and D.W. Masser 1980 for the non-CM case). *Let $\wp$ be a Weierstrass elliptic function with algebraic invariants g_2, g_3 and field of endomorphisms k. Let $u_1, \ldots, u_n$ be k-linearly independent complex numbers. Assume, for $1 \le i \le n$, that either $u_i \in \Omega$ or $\wp(u_i) \in \overline{\mathbb{Q}}$. Then the numbers $1, u_1, \ldots, u_n$ are linearly independent over the field $\overline{\mathbb{Q}}$.*

This means that *for an elliptic curve E that is defined over $\overline{\mathbb{Q}}$, if $u_1, \ldots, u_n$ are elements in $\mathcal{L}_E$ that are linearly independent over the field of endomorphisms of E, then the numbers $1, u_1, \ldots, u_n$ are linearly independent over $\overline{\mathbb{Q}}$.*

The method of Bertrand–Masser yields only weak Diophantine estimates (measures of linear independence of logarithms).

4.3 Further Results of Linear Independence

Theorem 26 deals only with periods and quasiperiods associated with one lattice; Theorem 27 deals only with elliptic logarithms of algebraic points on one elliptic curve. A far-reaching generalization of both results was achieved by G. Wüstholz in 1987 [249, 250, 251] when he succeeded in extending Baker's theorem to abelian varieties and integrals, and, more generally, to commutative algebraic groups. If we restrict his general result to products of a commutative linear group, of copies of elliptic curves, and of extensions of elliptic curves by the additive or the multiplicative group, the resulting statement settles the questions of linear independence of logarithms of algebraic numbers and of elliptic logarithms of algebraic points, including periods, quasiperiods, elliptic integrals of the first, second, or third kind. This is a main step toward an answer to the questions of M. Kontsevich and D. Zagier on periods [118].

Wüstholz's method can be extended to yield measures of linear independence of logarithms of algebraic points on an algebraic group. The first effective such lower bounds were given in 1989 [188, 189]. As a special case, they provide the first measures of linear independence for elliptic logarithms that is also valid in the non-CM case. More generally, they give effective lower bounds for any nonvanishing linear combination of logarithms of algebraic points on algebraic groups (including usual logarithms, elliptic logarithms, elliptic integrals of any kind).

Refinements were obtained by N. Hirata-Kohno [100, 101, 102, 103, 104], S. David [60], N. Hirata-Kohno and S. David [62], M. Ably [2, 3], and É. Gaudron [86, 87, 88], who uses not only Hirata's reduction argument, but also the work of J-B. Bost [33] (slope inequalities) involving Arakelov's theory. For instance, thanks to the recent work of David and Hirata-Kohno on the one hand, of Gaudron on the other, one knows that the above-mentioned nonvanishing linear combinations of logarithms of algebraic points are not Liouville numbers.

In the p-adic case there is a paper of G. Rémond and F. Urfels [194] dealing with two elliptic logarithms.

Further applications to elliptic curves of the Baker–Masser–Wüstholz method were derived by D.W. Masser and G. Wüstholz [163, 164].

A survey on questions related to the isogeny theorem is [178]. Other surveys dealing with the questions of *small points*, Bogomolov conjecture, and the André Oort conjecture are [59, 61]. We do not cover these aspects of the theory in the present paper. Other related topics that would deserve more attention are the theory of height and theta functions as well as ultrametric questions.

Extensions of the above-mentioned results to abelian varieties were considered by D.W. Masser [145, 146, 147, 148, 149, 150, 153, 155, 156, 157], S. Lang [123], J. Coates and S. Lang [56], D. Bertrand and Y.Z. Flicker [28], Y.Z. Flicker [84], D. Bertrand [25, 26]. For instance, J. Wolfart and G. Wüstholz [245] have shown that the only linear dependence relations with algebraic coefficients between the values $B(a, b)$ of the Euler beta function at points $(a, b) \in \mathbb{Q}^2$ are those that follow from the Deligne–Koblitz–Ogus relations (see further references in [243]).

5 Algebraic Independence of Numbers Related to Elliptic Functions

5.1 Small Transcendence Degree

We keep the notation and assumptions of Section 3.2.

The following extension of Theorem 24 to a result of algebraic independence containing Gel'fond's 1948 results on the exponential function (see Section 1.1) is a consequence of the work of many mathematicians, including A.O. Gel'fond [89], A.A. Šmelev [214, 215], W.D. Brownawell [35], W.D. Brownawell and K.K. Kubota [37], G. Wüstholz [246], D.W. Masser and G. Wüstholz [160], and others (further references are given in [35, 235, 236]).

Theorem 28.
1. *Assume* $(d_1 + d_2)m \geq 2(m + d_1 + 2d_2)$. *Then the field K_1 has transcendence degree ≥ 2 over $\mathbb{Q}$.*
2. *Assume* $(d_1 + d_2)m \geq m + 2(d_1 + 2d_2)$. *Then K_2 has transcendence degree ≥ 2 over $\mathbb{Q}$.*
3. *Assume* $(d_1 + d_2)m \geq 2m + d_1 + 2d_2$. *Then K_3 has transcendence degree ≥ 2 over $\mathbb{Q}$.*
4. *Assume* $(d_1 + d_2)m > m + d_1 + 2d_2$. *Then K_4 has transcendence degree ≥ 2 over $\mathbb{Q}$.*

Quantitative estimates (measures of algebraic independence) exist (R. Tubbs [220] and E.M. Jabbouri [105]).

Further related results are due to N.I. Fel'dman [81, 82], R. Tubbs [219, 220, 221, 222, 223, 224], É. Reyssat [201], M. Toyoda and T. Yasuda [218]. See also the measure of algebraic independence given by M. Ably in [1] and by S.O. Shestakov in [209].

A survey on results related to small transcendence degree is given in [236] (see also Chapter 13 of [177]).

Again, as for Theorem 24, there are extensions of Theorem 28 that include results on Weierstrass zeta functions as well as on functions of several variables, with a number of consequences related to abelian functions [237].

5.2 Algebraic Independence of Periods and Quasiperiods

In the 1970s, G.V. Chudnovsky proved strong results of algebraic independence (small transcendence degree) related to elliptic functions. One of his most spectacular contributions was obtained in 1976 [45] (see also [48] and [50]):

Theorem 29 (G.V. Chudnovsky, 1976). *Let $\wp$ be a Weierstrass elliptic function with invariants g_2, g_3. Let (ω_1, ω_2) be a basis of the lattice period of $\wp$ and $\eta_1 = \eta(\omega_1)$, $\eta_2 = \eta(\omega_2)$ the associated quasiperiods of the associated Weierstrass zeta function. Then at least two of the numbers g_2, g_3, ω_1, ω_2, η_1, η_2 are algebraically independent.*

A more precise result ([50] Chapter 7, Theorem 3.1) is that for any nonzero period ω, at least two of the four numbers g_2, g_3, ω/π, η/ω (with $\eta = \eta(\omega)$) are algebraically independent.

In the case that g_2 and g_3 are algebraic, one deduces from Theorem 29 that two among the four numbers ω_1, ω_2, η_1, η_2 are algebraically independent; this statement is also a consequence of the next result (Theorem 4 of [48]; see also [50, 235]):

Theorem 30 (G.V. Chudnovsky, 1981). *Assume that g_2 and g_3 are algebraic. Let ω be a nonzero period of $\wp$, set $\eta = \eta(\omega)$, and let u be a complex number that is not a period such that u and ω are $\mathbb{Q}$-linearly independent: $u \notin \mathbb{Q}\omega \cup \Omega$. Assume $\wp(u) \in \overline{\mathbb{Q}}$. Then the two numbers*

$$\zeta(u) - \frac{\eta}{\omega}u, \quad \frac{\eta}{\omega}$$

are algebraically independent.

From Theorem 29 or Theorem 30 one deduces the following result:

Corollary 31. *Let ω be a nonzero period of $\wp$ and $\eta = \eta(\omega)$. If g_2 and g_3 are algebraic, then the two numbers π/ω and η/ω are algebraically independent.*

The following consequence of Corollary 31 shows that in the CM case, Chudnovsky's results are sharp:

Corollary 32. *Assume that g_2 and g_3 are algebraic and the elliptic curve has complex multiplication. Let ω be a nonzero period of $\wp$. Then the two numbers ω and π are algebraically independent.*

As a consequence of formulas (5) and (6), one deduces the following corollary:

Corollary 33. *The numbers π and $\Gamma(1/4)$ are algebraically independent. Also the numbers π and $\Gamma(1/3)$ are algebraically independent.*

In connection with these results let us quote a conjecture of S. Lang from 1971 [121] p. 652.

Conjecture 34. If $j(\tau)$ is algebraic with $j'(\tau) \neq 0$, then $j'(\tau)$ is transcendental.

According to Siegel's relation (see [121] p. 652 and [66] Section 1.2.5 p. 165),

$$j'(\tau) = 18 \frac{\omega_1^2}{2\pi i} \cdot \frac{g_3}{g_2} \cdot j(\tau).$$

Conjecture 34 amounts to the transcendence of ω^2/π. Hence Corollary 32 implies that Conjecture 34 is true at least in the CM case (see [22]):

Corollary 35. *If $\tau \in \mathcal{H}$ is quadratic and $j'(\tau) \neq 0$, then π and $j'(\tau)$ are algebraically independent.*

A quantitative refinement (measure of algebraic independence) of Corollary 31 due to G. Philibert [181] turns out to be useful in connection with Nesterenko's work in 1996 (further references on this topic are given in [239]).

A transcendence measure for $\Gamma(1/4)$ was obtained by P. Philippon [186, 187] and S. Bruiltet [39]:

Theorem 36. *For $P \in \mathbb{Z}[X, Y]$ with degree d and height H,*

$$\log |P(\pi, \Gamma(1/4)| > -10^{326} \big((\log H + d \log(d + 1))d^2 (\log(d + 1)\big)^2.$$

Corollary 37. *The number $\Gamma(1/4)$ is not a Liouville number:*

$$\left| \Gamma(1/4) - \frac{p}{q} \right| > \frac{1}{q^{10^{330}}}.$$

Further algebraic independence results can be found in papers including those of D. Bertrand [20, 23], G.V. Chudnovsky [49] (however, see Zbl 0456.10016) and E. Reyssat [197, 200] (see also the Bourbaki lecture [231] and the book of E.B. Burger and R. Tubbs [42]). Among Chudnovsky's other contributions are results dealing with G-functions (see [50]; see also Y. André's work [7, 8]).

We conclude this section with the following open problem, which simultaneously generalizes Theorems 29 and 30 of G.V. Chudnovsky.

Conjecture 38. Let $\wp$ be a Weierstrass elliptic function with invariants g_2, g_3, let ω be a nonzero period of $\wp$, set $\eta = \eta(\omega)$, and let $u \in \mathbb{C} \setminus \{\mathbb{Q}\omega \cup \Omega\}$. Then at least two of the five numbers

$$g_2, \quad g_3, \quad \wp(u), \quad \zeta(u) - \frac{\eta}{\omega}u, \quad \frac{\eta}{\omega}$$

are algebraically independent.

Chudnovsky's method was extended by K.G. Vasil'ev [225] and P. Grinspan [94], who proved that at least two of the three numbers π, $\Gamma(1/5)$, and $\Gamma(2/5)$ are algebraically independent. Their proof involves the Jacobian of the Fermat curve $X^5 + Y^5 = Z^5$, which contains an abelian variety of dimension 2 as a factor. See also Pellarin's papers [179, 180].

5.3 Large Transcendence Degree

Another important (and earlier) contribution of G.V. Chudnovsky goes back to 1974, when he worked on extending Gel'fond's method in order to prove results on large transcendence degree (see references in [50, 231]).

Chudnovsky proved that three of the numbers

$$\alpha^\beta, \alpha^{\beta^2}, \ldots, \alpha^{\beta^{d-1}} \tag{11}$$

are algebraically independent if α is a nonzero algebraic number, $\log \alpha$ a nonzero logarithm of α, and β an algebraic number of degree $d \geq 7$. The same year, with a much more difficult and highly technical proof, he made the first substantial progress toward a proof that there exist at least n algebraically independent numbers in the set (11), provided that $d \geq 2^n - 1$. This was a remarkable achievement since no such result providing a lower bound for the transcendence degree was known (see [235] Section 2.1). Later, thanks to the work of several mathematicians, including P. Philippon (see [182] for his trick involving the introduction of redundant variables) and Yu. V. Nesterenko [166, 167, 168], the proof was completed and the exponential lower bound for d was reduced to a polynomial bound, until G. Diaz [63] obtained the best known results so far: the transcendence degree is at least $[(d + 1)/2]$.

During a short time, thanks to the work of Philippon, the elliptic results dealing with large transcendence degree where stronger than the exponential ones (see [235] p. 561).

Further results of algebraic independence related to elliptic functions are given in [46, 48, 50, 169, 170, 171, 197, 231].

In 1980, G.V. Chudnovsky [47] proved the Lindemann–Weierstrass theorem for $n = 2$ and $n = 3$ (small transcendence degree) by means of a clever variation of Gel'fond's method. At the same time he obtained the elliptic analogue in the CM case of the Lindemann–Weierstrass theorem for $n = 2$ and $n = 3$ in [46] and [47]. Also in [46] he announces further results of small transcendence degree (algebraic independence of four numbers).

This method was extended to large transcendence degree by P. Philippon [183, 184, 185] and G. Wüstholz [247, 248], who also succeeded in 1982 to prove the elliptic analogue of the Lindemann–Weierstrass theorem on the algebraic independence of $e^{\alpha_1}, \ldots, e^{\alpha_n}$ in the CM case:

Theorem 39. *Let $\wp$ be a Weierstrass elliptic function with algebraic invariants g_2, g_3 and complex multiplication. Let $\alpha_1, \ldots, \alpha_m$ be algebraic numbers that are linearly independent over the field of endomorphisms of E. Then the numbers $\wp(\alpha_1), \ldots, \wp(\alpha_n)$ are algebraically independent.*

The same conclusion should also hold in the non-CM case; so far, only the algebraic independence of at least $n/2$ of these numbers is known.

Further results on large transcendence degree are due to D.W. Masser and G. Wüstholz [161, 162], W.D. Brownawell [36], W.D. Brownawell and R. Tubbs [38], M. Takeuchi [216].

A survey on algebraic independence was written in 1979 by W.D. Brownawell [35]. The period prior to 1984 is covered by [235] (see also [236]), while [242] gives references for the period 1984–1997. A more recent reference is [177] Chapter 14.

5.4 Modular Functions and Ramanujan Functions

Ramanujan [193] introduced the following functions:

$$P(q) = 1 - 24 \sum_{n=1}^{\infty} \frac{nq^n}{1 - q^n}, \quad Q(q) = 1 + 240 \sum_{n=1}^{\infty} \frac{n^3 q^n}{1 - q^n},$$

$$R(q) = 1 - 504 \sum_{n=1}^{\infty} \frac{n^5 q^n}{1 - q^n}.$$

They are special cases of Fourier expansions of Eisenstein series. Recall the Bernoulli numbers B_k defined by

$$\frac{z}{e^z - 1} = 1 - \frac{z}{2} + \sum_{k=1}^{\infty} (-1)^{k+1} B_k \frac{z^{2k}}{(2k)!},$$

$$B_1 = 1/6, \quad B_2 = 1/30, \quad B_3 = 1/42.$$

For $k \geq 1$ the normalized Eisenstein series of weight k is ([116] Section 3.2, Proposition 6, [208] Section 7.4.2)

$$E_{2k}(q) = 1 + (-1)^k \frac{4k}{B_k} \sum_{n=1}^{\infty} \frac{n^{2k-1} q^n}{1 - q^n}.$$

The connection with (4) is

$$E_{2k}(q) = \frac{1}{2\zeta(2k)} \cdot G_k(\tau),$$

for $k \geq 2$, where $q = e^{2\pi i \tau}$ ([48] Section 3.2, Proposition 6). In particular,

$$G_2(\tau) = \frac{\pi^4}{3^2 \cdot 5} \cdot E_4(q), \qquad G_3(\tau) = \frac{2\pi^6}{3^3 \cdot 5 \cdot 7} \cdot E_6(q).$$

With Ramanujan's notation we have

$$P(q) = E_2(q), \quad Q(q) = E_4(q), \quad R(q) = E_6(q).$$

The discriminant Δ and the modular invariant J are related to these functions by Jacobi's product formula ([127] Section 18.4 and [208] Sections 7.2.3, 7.3.3, 7.4.4)

$$\Delta = \frac{(2\pi)^{12}}{12^3} \cdot (Q^3 - R^2) = (2\pi)^{12} q \prod_{n=1}^{\infty} (1 - q^n)^{24},$$

and

$$J = \frac{(2\pi)^{12} Q^3}{\Delta} = \frac{(2^4 3^2 5 G_2)^3}{\Delta}.$$

Let q be a complex number, $0 < |q| < 1$. There exists τ in the upper half-plane $\mathcal{H}$ such that $q = e^{2\pi i \tau}$. Select any twelfth root ω of $\Delta(q)$. The invariants g_2 and g_3 of the Weierstrass $\wp$ function attached to the lattice $(\mathbb{Z} + \mathbb{Z}\tau)\omega$ satisfy $g_2^3 - 27g_3^2 = 1$ and (see [66] Section 1.2.2 p. 163, [127], Section 4.2, Proposition 4 and Section 18.3)

$$P(q) = 3\frac{\omega}{\pi} \cdot \frac{\eta}{\pi}, \quad Q(q) = \frac{3}{4}\left(\frac{\omega}{\pi}\right)^4 g_2, \quad R(q) = \frac{27}{8}\left(\frac{\omega}{\pi}\right)^6 g_3.$$

According to formulas (5) and (6), here are a few special values (see, for instance, [4], [177] Section 3.1 and [241]).

– For $\tau = i, q = e^{-2\pi}$,

$$P(e^{-2\pi}) = \frac{3}{\pi}, \quad Q(e^{-2\pi}) = 3\left(\frac{\omega_1}{\pi}\right)^4, \quad R(e^{-2\pi}) = 0, \quad \text{and}$$

$$\Delta(e^{-2\pi}) = 2^6 \omega_1^{12}, \tag{12}$$

with

$$\omega_1 = \frac{\Gamma(1/4)^2}{\sqrt{8\pi}} = 2.6220575542\ldots.$$

– For $\tau = \varrho, q = -e^{-\pi\sqrt{3}}$,

$$P(-e^{-\pi\sqrt{3}}) = \frac{2\sqrt{3}}{\pi}, \quad Q(-e^{-\pi\sqrt{3}}) = 0, \quad R(-e^{-\pi\sqrt{3}}) = \frac{27}{2}\left(\frac{\omega_1}{\pi}\right)^6,$$

$$\Delta(-e^{-\pi\sqrt{3}}) = -2^4 3^3 \omega_1^{12}, \tag{13}$$

with

$$\omega_1 = \frac{\Gamma(1/3)^3}{2^{4/3}\pi} = 2.428650648\ldots.$$

5.5 Mahler–Manin Problem on $J(q)$

After Schneider's theorem (Corollary 21) on the transcendence of the values of the modular function $j(\tau)$, the first results on Eisenstein series (cf. Section 5.6) go back to 1977 with D. Bertrand's work [21, 19]. See also his papers [18, 20, 23, 24], his work [29] with M. Laurent on values of theta functions, and Yanchenko's paper [252]. Further related results are Theorems 5 and 6 (p. 344) and Theorem 4 (p. 347) in Chudnovsky's lecture at the Helsinki ICM in 1978 [48].

The first transcendence proof using modular forms is due to a team from St Étienne (K. Barré-Sirieix, G. Diaz, F. Gramain and G. Philibert), whence the nickname *théorème stéphanois* for the next result; see [16] (see also [91, 92, 90] and Chapter 2 of [177]). Theorem 40 answers a conjecture of K. Mahler [139, 140] in the complex case and of Yu. V. Manin [142] in the *p*-adic case. Manin's question on the arithmetic nature of the *p*-adic number $J(q)$ is motivated by Mazur's theory, but he also asked "an obvious analogue" in the complex case; see Conjecture 43 below). We state the result only in the complex case; the paper [16] solves both cases.

Theorem 40 (K. Barré, G. Diaz, F. Gramain, G. Philibert, 1996). *Let* $q \in \mathbb{C}$, $0 < |q| < 1$. *If q is algebraic, then $J(q)$ is transcendental.*

The solution of Manin's problem in the *p*-adic case has several consequences. It is a tool both for R. Greenberg in his study of zeros of *p*-adic L functions and for H. Hida, J. Tilouine, and É. Urban in their solution of the main conjecture for the Selmer group of the symmetric square of an elliptic curve with multiplicative reduction at *p* (references are given in [239]).

The proof of Theorem 40 involves upper bounds for the growth of the coefficients of the modular function $J(q)$. Such estimates were produced first by K. Mahler [141] Section 3. A refined estimate, due to N. Brisebarre and G. Philibert [34], for the coefficients $c_k(m)$ (which are nonnegative rational integers) in

$$\left(qJ(q)\right)^k = \sum_{m=0}^{\infty} c_k(m)q^m$$

is

$$c_k(m) \leq e^{4\sqrt{km}}.$$

According to a remark by D. Bertrand (Lemma 1 in [241] and Lemma 2.4 p. 17 in [177]; see also Lemma 2 in [15] and Lemma 1 in [27]), the upper bound

$$|\tilde{c}_{N,k}(m)| \leq C^N m^{12N}$$

($0 \leq k \leq N$, $N \geq 1$, $m \geq 1$, with an absolute constant C) for the coefficients in the Taylor development at the origin of $\Delta^{2N} J^k$,

$$\Delta(q)^{2N} J(q)^k = \sum_{m=1}^{\infty} \tilde{c}_{Nk}(m)q^m,$$

is sufficient for the proof of Theorem 40 and is an easy consequence of a theorem of Hecke ([208] Section 7.4.3, Theorem 5), together with the fact that Δ^2 and $\Delta^2 J$ are parabolic modular forms of weight 24.

One of the main tools involved in the proof of Theorem 40 is an estimate for the degrees and height of $J(q^n)$ in terms of $J(q)$ (which is assumed to be algebraic) and $n \geq 1$. There exists a symmetric polynomial $\Phi_n \in \mathbb{Z}[X, Y]$, of degree

$$\psi(n) = n \prod_{p|n} \left(1 + \frac{1}{p}\right)$$

in each variable, such that $\Phi_n(J(q), J(q^n)) = 0$. Again, K. Mahler was the first to investigate the coefficients of the polynomial $\Phi_n(X, Y)$: in [141] he proved that its length (sum of the absolute values of the coefficients) satisfies

$$L(\Phi_n) \leq e^{cn^{3/2}}$$

with an absolute constant c. In the special case $n = 2^m$ he had an earlier stronger result in [140], namely

$$L(\Phi_n) \leq 2^{57n} n^{36n},$$

and he claimed (see [140] p. 97) that if the sharper upper bound

$$L(\Phi_n) \leq 2^{Cn}$$

with a positive absolute constant $C > 0$ were true for $n = 2^m$, he could prove Theorem 40. However, in 1984, P. Cohen [58] produced asymptotic estimates that show that Mahler's expectation was too optimistic:

$$\lim_{\substack{n=2^m \\ m \to \infty}} \frac{1}{n \log n} \log L(\Phi_n) = 9.$$

In fact she proved more precise results, without the condition $n = 2^m$, which imply, for instance, $\log L(\Phi_n) \sim 6\psi(n) \log n$ for $n \to \infty$.

Further related results are given in [67] (G. Diaz and G. Philibert) for the j-function and [159] (D.W. Masser) for the $\wp$-function.

The proof of [16] can be adapted to yield quantitative estimates [14, 15].

A reformulation of Theorem 40 on the transcendence of $J(q)$ is the following mixed analogue of the four exponentials conjecture (Conjecture 12):

Corollary 41. *Let* $\log \alpha$ *be a logarithm of a nonzero algebraic number. Let* $\mathbb{Z}\omega_1 + \mathbb{Z}\omega_2$ *be a lattice with algebraic invariants* g_2, g_3. *Then the determinant*

$$\begin{vmatrix} \omega_1 & \log \alpha \\ \omega_2 & 2\pi i \end{vmatrix}$$

does not vanish.

The four exponentials conjecture for the product of an elliptic curve by the multiplicative group is the following more general open problem:

Conjecture 42. Let $\wp$ be a Weierstrass elliptic function with algebraic invariants g_2, g_3. Let E be the corresponding elliptic curve, u_1 and u_2 two elements in $\mathcal{L}_E$, and $\log \alpha_1$, $\log \alpha_2$ two logarithms of algebraic numbers. Assume further that the two rows of the matrix

$$M = \begin{pmatrix} u_1 & \log \alpha_1 \\ u_2 & \log \alpha_2 \end{pmatrix}$$

are linearly independent over $\mathbb{Q}$. Then the determinant of M does not vanish.

Another special case of Conjecture 42, stronger than Corollary 41, is the next question of Yu. V. Manin, who asks in Section 4.2 of [142] to determine the nature of the invariant of the complex elliptic curve having periods 1 and a quotient $(\log \alpha_1)/(\log \alpha_2)$ of two logarithms of algebraic numbers:

Conjecture 43 (Yu.V. Manin). Let $\log \alpha_1$ and $\log \alpha_2$ be two nonzero logarithms of algebraic numbers and let $\mathbb{Z}\omega_1 + \mathbb{Z}\omega_2$ be a lattice with algebraic invariants g_2 and g_3. Then

$$\frac{\omega_1}{\omega_2} \neq \frac{\log \alpha_1}{\log \alpha_2}.$$

In this direction let us quote some of the open problems raised by G. Diaz [65, 66].

Conjecture 44 (G. Diaz).
1. For any $z \in \mathbb{C}$ with $|z| = 1$ and $z \neq \pm 1$, the number $e^{2\pi i z}$ is transcendental.
2. If q is an algebraic number with $0 < |q| < 1$ such that $J(q) \in [0, 1728]$, then $q \in \mathbb{R}$.
3. The function J is injective on the set of algebraic numbers α with $0 < |\alpha| < 1$.

Remark (G. Diaz). Part 3 of Conjecture 44 implies the other two and also follows from the four exponentials conjecture. It also follows from the next conjecture of D. Bertrand.

Conjecture 45 (D. Bertrand). If α_1 and α_2 are two multiplicatively independent algebraic numbers in the domain $\{q \in \mathbb{C}; 0 < |q| < 1\}$, then the two numbers $J(\alpha_1)$ and $J(\alpha_2)$ are algebraically independent.

Conjecture 45 (see [27], where Section 5 is devoted to conjectural statements inspired by a conjecture of Oort and André) implies the special case of the four exponentials conjecture, where two of the algebraic numbers are roots of unity and the two others have modulus $\neq 1$.

5.6 Nesterenko's Theorem

In 1976 [18], D. Bertrand pointed out that Schneider's theorem on the transcendence of ω/π implies the following statement:

For any $q \in \mathbb{C}$ with $0 < |q| < 1$, at least one of the two numbers $Q(q)$, $R(q)$ is transcendental.

He also proved in [18] the p-adic analogue by means of a new version of the Schneider–Lang criterion for meromorphic functions (he allows one essential singularity), which he applied to Jacobi–Tate elliptic functions (see also [252]). Two years later [20], he noticed that Theorem 29 yields the following:

For any $q \in \mathbb{C}$ with $0 < |q| < 1$, at least two of the numbers $P(q)$, $Q(q)$, $R(q)$ are algebraically independent.

The following result of Yu. V. Nesterenko [172, 173] (see also [175, 176, 239, 241, 32] as well as Chapters 3 and 4 of [177]) goes one step further:

Theorem 46 (Nesterenko, 1996). *For any $q \in \mathbb{C}$ with $0 < |q| < 1$, three of the four numbers q, $P(q)$, $Q(q)$, $R(q)$ are algebraically independent.*

Among the tools used by Nesterenko in his proof is the following result due to K. Mahler [138] (see also Chapter 1 of [177]):

The functions P, Q, R are algebraically independent over $\mathbb{C}(q)$.

Also he uses the fact that they satisfy a system of differential equations for $D = q \, d/dq$ discovered by S. Ramanujan in 1916 [193] (see also Chapters 1 and 3 of [177]):

$$12\frac{DP}{P} = P - \frac{Q}{P}, \quad 3\frac{DQ}{Q} = P - \frac{R}{Q}, \quad 2\frac{DR}{R} = P - \frac{Q^2}{R}.$$

One of the main steps in his original proof [172, 173] is the following zero estimate:

Theorem 47 (Nesterenko's zero estimate). *Let L_0 and L be positive integers, $A \in \mathbb{C}[q, X_1, X_2, X_3]$ a nonzero polynomial in four variables of degree $\leq L_0$ in q and $\leq L$ in each of the three other variables X_1, X_2, X_3. Then the multiplicity at the origin of the analytic function $A\big(q, P(q), Q(q), R(q)\big)$ is at most $2 \cdot 10^{45} L_0 L^3$.*

In the special case in which $J(q)$ is algebraic, P. Philippon [187] produced an alternative proof for Nesterenko's result in which this zero estimate is not used; instead of it, he used Philibert's measure of algebraic independence for ω/π and η/π (see [181] and Section 5.2 above). However, Philibert's proof requires a zero estimate for algebraic groups.

Using (12) one deduces from Theorem 46 (see [177] Section 3.1, Corollary 1.2) the following corollary:

Corollary 48. *The three numbers π, e^{π}, $\Gamma(1/4)$ are algebraically independent.*

Using (13) one deduces (see [177] Section 1.3.1, Corollary 3.2, Remark (ii)) the following:

Corollary 49. *The three numbers π, $e^{\pi\sqrt{3}}$, $\Gamma(1/3)$ are algebraically independent.*

Consequences of Corollary 48 are the transcendence of the numbers

$$\sigma_{\mathbb{Z}[i]}(1/2) = 2^{5/4}\pi^{1/2}e^{\pi/8}\Gamma(1/4)^{-2}$$

and (P. Bundschuh [41])

$$\sum_{n=0}^{\infty} \frac{1}{n^2+1} = \frac{1}{2} + \frac{\pi}{2} \cdot \frac{e^\pi + e^{-\pi}}{e^\pi - e^{-\pi}}.$$

D. Duverney, K. and K. Nishioka, and I. Shiokawa [68, 69, 70, 71, 72, 73] as well as D. Bertrand [27] derived from Nesterenko's theorem a number of interesting corollaries, including the following ones ([177] Chapter 3).

Corollary 50. *The Rogers–Ramanujan continued fraction*

$$RR(\alpha) = 1 + \cfrac{\alpha}{1 + \cfrac{\alpha^2}{1 + \cfrac{\alpha^3}{1 + \cfrac{}{\ddots}}}}$$

is transcendental for any algebraic α with $0 < |\alpha| < 1$.

Corollary 51. *Let $(F_n)_{n\geq 0}$ be the Fibonacci sequence: $F_0 = 0$, $F_1 = 1$, $F_n = F_{n-1} + F_{n-2}$. Then the number*

$$\sum_{n=1}^{\infty} \frac{1}{F_n^2}$$

is transcendental.

Jacobi theta series ([43] Chapter V, [244] Chapter XXI and [177] Section 3.1.3) are defined by

$$\theta_2(q) = 2q^{1/4} \sum_{n\geq 0} q^{n(n+1)} = 2q^{1/4} \prod_{n=1}^{\infty} (1 - q^{4n})(1 + q^{2n}),$$

$$\theta_3(q) = \sum_{n\in\mathbb{Z}} q^{n^2} = \prod_{n=1}^{\infty} (1 - q^{2n})(1 + q^{2n-1})^2,$$

$$\theta_4(q) = \theta_3(-q) = \sum_{n\in\mathbb{Z}} (-1)^n q^{n^2} = \prod_{n=1}^{\infty} (1 - q^{2n})(1 - q^{2n-1})^2.$$

Corollary 52. *Let i, j and $k \in \{2, 3, 4\}$ with $i \neq j$. Let $q \in \mathbb{C}$ satisfy $0 < |q| < 1$. Then each of the two fields*

$$\mathbb{Q}(q, \theta_i(q), \theta_j(q), D\theta_k(q)) \quad \text{and} \quad \mathbb{Q}(q, \theta_k(q), D\theta_k(q), D^2\theta_k(q))$$

has transcendence degree ≥ 3 over $\mathbb{Q}$.

As an example, *for an algebraic number $q \in \mathbb{C}$ with $0 < |q| < 1$, the three numbers*

$$\sum_{n \geq 0} q^{n^2}, \quad \sum_{n \geq 1} n^2 q^{n^2}, \quad \sum_{n \geq 1} n^4 q^{n^2}$$

are algebraically independent. In particular, *the number*

$$\theta_3(q) = \sum_{n \in \mathbb{Z}} q^{n^2}$$

is transcendental. The number $\theta_3(q)$ was explicitly considered by Liouville as far back as 1851 (see [174] p. 295 and [177] p. 30).

The proof by Yu. V. Nesterenko is effective and yields quantitative refinements (measures of algebraic independence): [93, 174, 187].

5.7 Further Open Problems

Among many open problems, we mention

- the algebraic independence of the three numbers π, $\Gamma(1/3)$, $\Gamma(1/4)$.
- the algebraic independence of three numbers among π, $\Gamma(1/5)$, $\Gamma(2/5)$, $e^{\pi\sqrt{5}}$.
- the algebraic independence of the four numbers e, π, e^{π} and $\Gamma(1/4)$.

The main conjectures in this domain are due to S. Schanuel, A. Grothendieck, Y. André [9], and C. Bertolin [17]. Chudnovsky's proof of the algebraic independence of π and $\Gamma(1/4)$ involves elliptic functions; Nesterenko's proof of the algebraic independence of π and e^{π} requires modular functions. One may expect that higher-dimensional objects (abelian varieties, motives) may be required in order to go further. In this respect we conclude by alluding to the remarkable progress that has been achieved recently in finite characteristic (after the work by Jing Yu, G.W. Anderson and D. Thakur, L. Denis, W.D. Brownawell, J.F. Voloch, M. Papanikolas, among others).

References

1. M. ABLY – "Résultats quantitatifs d'indépendance algébrique pour les groupes algébriques," *J. Number Theory* **42** (1992), no. 2, pp. 194–231.
2. — , "Formes linéaires de logarithmes de points algébriques sur une courbe elliptique de type CM," *Ann. Inst. Fourier (Grenoble)* **50** (2000), no. 1, pp. 1–33.
3. M. ABLY & É. GAUDRON – "Approximation diophantienne sur les courbes elliptiques à multiplication complexe," *C. R. Math. Acad. Sci. Paris* **337** (2003), no. 10, pp. 629–634.
4. M. ABRAMOWITZ & I. A. STEGUN – *Handbook of mathematical functions with formulas, graphs, and mathematical tables*, A Wiley-Interscience Publication, New York: John Wiley & Sons, Inc; Washington, D.C, 1984, Reprint of the 1972 ed.
5. M. ANDERSON – "Inhomogeneous linear forms in algebraic points of an elliptic function," in *Transcendence theory: advances and applications (Proc. Conf., Univ. Cambridge, Cambridge, 1976)*, Academic Press, London, 1977, pp. 121–143.

6. M. ANDERSON & D. W. MASSER – "Lower bounds for heights on elliptic curves," *Math. Z.* **174** (1980), no. 1, pp. 23–34.

7. Y. ANDRÉ – *G-functions and geometry*, Aspects of Mathematics, E13, Friedr. Vieweg & Sohn, Braunschweig, 1989.

8. ⸻ , "*G*-fonctions et transcendance," *J. reine angew. Math.* **476** (1996), pp. 95–125.

9. ⸻ , *Une introduction aux motifs (motifs purs, motifs mixtes, périodes)*, Panoramas et Synthèses, vol. 17, Société Mathématique de France, Paris, 2004.

10. A. BAKER – "On the quasi-periods of the Weierstrass ζ-function," *Nachr. Akad. Wiss. Göttingen Math.-Phys. Kl. II* **1969** (1969), pp. 145–157.

11. ⸻ , "An estimate for the $\wp$-function at an algebraic point," *Amer. J. Math.* **92** (1970), pp. 619–622.

12. ⸻ , "On the periods of the Weierstrass $\wp$-function," in *Symposia Mathematica, Vol. IV (INDAM, Rome, 1968/69)*, Academic Press, London, 1970, pp. 155–174.

13. ⸻ , *Transcendental number theory*, Cambridge University Press, London, 1975.

14. K. BARRÉ – "Mesures de transcendance pour l'invariant modulaire," *C. R. Acad. Sci. Paris Sér. I Math.* **323** (1996), no. 5, pp. 447–452.

15. ⸻ , "Mesure d'approximation simultanée de q et $J(q)$," *J. Number Theory* **66** (1997), no. 1, pp. 102–128.

16. K. BARRÉ-SIRIEIX, G. DIAZ, F. GRAMAIN & G. PHILIBERT – "Une preuve de la conjecture de Mahler-Manin," *Invent. Math.* **124** (1996), no. 1–3, pp. 1–9.

17. C. BERTOLIN – "Périodes de 1-motifs et transcendance," *J. Number Theory* **97** (2002), no. 2, pp. 204–221.

18. D. BERTRAND – "Séries d'Eisenstein et transcendance," *Bull. Soc. Math. France* **104** (1976), no. 3, pp. 309–321.

19. ⸻ , "Transcendance de valeurs de la fonction gamma d'après G. V. Chudnovsky (Dokl. Akad. Nauk Ukrain. SSR Ser. A **1976**, no. 8, 698–701)," in *Séminaire Delange-Pisot-Poitou, 17e année (1975/76), Théorie des nombres: Fasc. 2, Exp. No. G8*, Secrétariat Math., Paris, 1977, p. 5.

20. ⸻ , "Fonctions modulaires, courbes de Tate et indépendance algébrique," in *Séminaire Delange-Pisot-Poitou, 19e année: 1977/78, Théorie des nombres, Fasc. 2*, Secrétariat Math., Paris, 1978, p. Exp. No. 36, 11.

21. ⸻ , "Modular function and algebraic independence," in *Proceedings of the Conference on p-adic Analysis (Nijmegen, 1978)*, Report, vol. 7806, Katholieke Univ. Nijmegen, 1978, pp. 16–23.

22. ⸻ , "Propriétés arithmétiques des dérivées de la fonction modulaire $j(\tau)$," in *Séminaire de Théorie des Nombres 1977–1978*, CNRS, Talence, 1978, p. Exp. No. 22, 4.

23. ⸻ , "Fonctions modulaires et indépendance algébrique. II," in *Journées Arithmétiques de Luminy (Colloq. Internat. CNRS, Centre Univ. Luminy, Luminy, 1978)*, Astérisque, vol. 61, Soc. Math. France, Paris, 1979, pp. 29–34.

24. ⸻ , "Sur les périodes de formes modulaires," *C. R. Acad. Sci. Paris Sér. A-B* **288** (1979), no. 10, pp. A531–A534.

25. ⸻ , "Variétés abeliennes et formes linéaires d'intégrales elliptiques," in *Théorie des nombres, Sémin. Delange-Pisot-Poitou, Paris 1979–80*, Prog. Math. 12, 15–27, 1981.

26. ⸻ , "Endomorphismes de groupes algébriques; applications arithmétiques," in *Diophantine approximations and transcendental numbers (Luminy, 1982)*, Progr. Math., vol. 31, Birkhäuser Boston, Boston, MA, 1983, pp. 1–45.

27. ⸻ , "Theta functions and transcendence," *Ramanujan J.* **1** (1997), no. 4, pp. 339–350, International Symposium on Number Theory (Madras, 1996).

28. D. BERTRAND & Y. Z. FLICKER – "Linear forms on abelian varieties over local fields," *Acta Arith.* **38** (1980/81), no. 1, pp. 47–61.

29. D. BERTRAND & M. LAURENT – "Propriétés de transcendance de nombres liés aux fonctions thêta," *C. R. Acad. Sci. Paris Sér. I Math.* **292** (1981), no. 16, pp. 747–749.

30. D. BERTRAND & D. MASSER – "Formes linéaires d'intégrales abéliennes," *C. R. Acad. Sci. Paris Sér. A-B* **290** (1980), no. 16, pp. A725–A727.

31. —, "Linear forms in elliptic integrals," *Invent. Math.* **58** (1980), no. 3, pp. 283–288.

32. V. BOSSER – "Indépendance algébrique de valeurs de séries d'Eisenstein (théorème de Nesterenko)," in *Formes modulaires et transcendance*, Sémin. Congr., vol. 12, Soc. Math. France, Paris, 2005, pp. 119–178.

33. J.-B. BOST – "Périodes et isogénies des variétés abéliennes sur les corps de nombres (d'après D. Masser et G. Wüstholz)," *Astérisque* **237** (1996), no. 4, pp. 115–161, Séminaire Bourbaki, Vol. 1994/95, Exp. No. 795.

34. N. BRISEBARRE & G. PHILIBERT – "Effective lower and upper bounds for the Fourier coefficients of powers of the modular invariant j," *Journal of the Ramanujan Mathematical Society* **20** (2005), no. 4, pp. 255–282.

35. W. D. BROWNAWELL – "On the development of Gel'fond's method," in *Number theory, Carbondale 1979 (Proc. Southern Illinois Conf., Southern Illinois Univ., Carbondale, Ill., 1979)*, Lecture Notes in Math., vol. 751, Springer, Berlin, 1979, pp. 18–44.

36. —, "Large transcendence degree revisited. I. Exponential and non-CM cases," in *Diophantine approximation and transcendence theory (Bonn, 1985)*, Lecture Notes in Math., vol. 1290, Springer, Berlin, 1987, pp. 149–173.

37. W. D. BROWNAWELL & K. K. KUBOTA – "The algebraic independence of Weierstrass functions and some related numbers," *Acta Arith.* **33** (1977), no. 2, pp. 111–149.

38. W. D. BROWNAWELL & R. TUBBS – "Large transcendence degree revisited. II. The CM case," in *Diophantine approximation and transcendence theory (Bonn, 1985)*, Lecture Notes in Math., vol. 1290, Springer, Berlin, 1987, pp. 175–188.

39. S. BRUILTET – "D'une mesure d'approximation simultanée à une mesure d'irrationalité: le cas de $\Gamma(1/4)$ et $\Gamma(1/3)$," *Acta Arith.* **104** (2002), no. 3, pp. 243–281.

40. P. BUNDSCHUH – "Ein Approximationsmass für transzendente Lösungen gewisser transzendenter Gleichungen," *J. reine angew. Math.* **251** (1971), pp. 32–53.

41. —, "Zwei Bemerkungen über transzendente Zahlen," *Monatsh. Math.* **88** (1979), no. 4, pp. 293–304.

42. E. B. BURGER & R. TUBBS – *Making transcendence transparent*, Springer-Verlag, New York, 2004, An intuitive approach to classical transcendental number theory.

43. K. CHANDRASEKHARAN – *Elliptic functions*, Grundlehren der Mathematischen Wissenschaften, vol. 281, Springer-Verlag, Berlin, 1985.

44. S. CHOWLA & A. SELBERG – "On Epstein's zeta-function," *J. reine angew. Math.* **227** (1967), pp. 86–110.

45. G. V. CHUDNOVSKY – "Algebraic independence of constants connected with the exponential and the elliptic functions," *Dokl. Akad. Nauk Ukrain. SSR Ser. A* **8** (1976), pp. 698–701, 767.

46. —, "Indépendance algébrique des valeurs d'une fonction elliptique en des points algébriques. Formulation des résultats," *C. R. Acad. Sci. Paris Sér. A-B* **288** (1979), no. 8, pp. A439–A440.

47. —, "Algebraic independence of the values of elliptic function at algebraic points," *Invent. Math.* **61** (1980), no. 3, pp. 267–290, Elliptic analogue of the Lindemann-Weierstrass theorem.

48. —, "Algebraic independence of values of exponential and elliptic functions," in *Proceedings of the International Congress of Mathematicians (Helsinki, 1978)* (Helsinki), Acad. Sci. Fennica, 1980, pp. 339–350.

49. — , "Indépendance algébrique dans la méthode de Gelfond-Schneider," *C. R. Acad. Sci., Paris, Sér. A* **291** (1980), pp. 365–368, *see* Zbl 0456.10016.
50. — , *Contributions to the theory of transcendental numbers*, Mathematical Surveys and Monographs, vol. 19, American Mathematical Society, Providence, RI, 1984.
51. J. COATES – "An application of the division theory of elliptic functions to Diophantine approximation," *Invent. Math.* **11** (1970), pp. 167–182.
52. — , "An application of the Thue-Siegel-Roth theorem to elliptic functions," *Proc. Cambridge Philos. Soc.* **69** (1971), pp. 157–161.
53. — , "Linear forms in the periods of the exponential and elliptic functions," *Invent. Math.* **12** (1971), pp. 290–299.
54. — , "The transcendence of linear forms in $\omega_1, \omega_2, \eta_1, \eta_2, 2\pi i$," *Amer. J. Math.* **93** (1971), pp. 385–397.
55. — , "Linear relations between $2\pi i$ and the periods of two elliptic curves," in *Diophantine approximation and its applications (Proc. Conf., Washington, D.C., 1972)*, Academic Press, New York, 1973, pp. 77–99.
56. J. COATES & S. LANG – "Diophantine approximation on Abelian varieties with complex multiplication," *Invent. Math.* **34** (1976), no. 2, pp. 129–133 (= [130] pp. 236–240).
57. H. COHEN – "Elliptic curves," in *From Number Theory to Physics (Les Houches, 1989)*, Springer, Berlin, 1992, pp. 212–237.
58. P. B. COHEN – "On the coefficients of the transformation polynomials for the elliptic modular function," *Math. Proc. Cambridge Philos. Soc.* **95** (1984), no. 3, pp. 389–402.
59. — , "Perspectives de l'approximation diophantienne et de la transcendance," *Ramanujan J.* **7** (2003), no. 1–3, pp. 367–384, Rankin memorial issues.
60. S. DAVID – "Minorations de formes linéaires de logarithmes elliptiques," *Mém. Soc. Math. France (N.S.)* **62** (1995), pp. 1–143.
61. — , "On the height of subvarieties of group varieties," (preprint).
62. S. DAVID & N. HIRATA-KOHNO – "Recent progress on linear forms in elliptic logarithms," in *A panorama of number theory or the view from Baker's garden (Zürich, 1999)*, Cambridge Univ. Press, Cambridge, 2002, pp. 26–37.
63. G. DIAZ – "Grands degrés de transcendance pour des familles d'exponentielles," *J. Number Theory* **31** (1989), no. 1, pp. 1–23.
64. — , "Minorations de combinaisons linéaires non homogènes pour un logarithme elliptique," *C. R. Acad. Sci. Paris Sér. I Math.* **318** (1994), no. 10, pp. 879–883.
65. — , "La conjecture des quatre exponentielles et les conjectures de D. Bertrand sur la fonction modulaire," *J. Théor. Nombres Bordeaux* **9** (1997), no. 1, pp. 229–245.
66. — , "Transcendance et indépendance algébrique: liens entre les points de vue elliptique et modulaire," *Ramanujan J.* **4** (2000), no. 2, pp. 157–199.
67. G. DIAZ & G. PHILIBERT – "Growth properties of the modular function j," *J. Math. Anal. Appl.* **139** (1989), no. 2, pp. 382–389.
68. D. DUVERNEY, K. NISHIOKA, K. NISHIOKA & I. SHIOKAWA – "Transcendence of Jacobi's theta series," *Proc. Japan Acad. Ser. A Math. Sci.* **72** (1996), no. 9, pp. 202–203.
69. — , "Transcendence of Rogers-Ramanujan continued fraction and reciprocal sums of Fibonacci numbers," *Proc. Japan Acad. Ser. A Math. Sci.* **73** (1997), no. 7, pp. 140–142.
70. — , "Transcendence of Jacobi's theta series and related results," in *Number theory (Eger, 1996)*, W. de Gruyter, Berlin, 1998, pp. 157–168.
71. — , "Transcendence of Rogers-Ramanujan continued fraction and reciprocal sums of Fibonacci numbers," *Sūrikaisekikenkyūsho Kōkyūroku* **1060** (1998), pp. 91–100, Number theory and its applications (Japanese) (Kyoto, 1997).

72. D. DUVERNEY & I. SHIOKAWA – "On some arithmetical properties of Rogers-Ramanujan continued fraction," in *Colloque Franco-Japonais: Théorie des Nombres Transcendants (Tokyo, 1998)*, Sem. Math. Sci., vol. 27, Keio Univ., Yokohama, 1999, pp. 91–100.

73. — , "On some arithmetical properties of Rogers-Ramanujan continued fraction," *Osaka J. Math.* **37** (2000), no. 3, pp. 759–771.

74. A. FAISANT & G. PHILIBERT – "Mesure d'approximation simultanée pour la fonction modulaire *j* et résultats connexes," in *Séminaire de théorie des nombres, Paris 1984–85*, Progr. Math., vol. 63, Birkhäuser Boston, Boston, MA, 1986, pp. 67–78.

75. — , "Quelques résultats de transcendance liés à l'invariant modulaire *j*," *J. Number Theory* **25** (1987), no. 2, pp. 184–200.

76. N. I. FEL′DMAN – "On the measure of transcendency of the logarithms of algebraic numbers and elliptic constants," *Uspehi Matem. Nauk (N.S.)* **4** (1949), no. 1(29), p. 190.

77. — , "The approximation of certain transcendental numbers. II. The approximation of certain numbers connected with the Weierstrass function $\wp(z)$," *Izvestiya Akad. Nauk SSSR. Ser. Mat.* **15** (1951), pp. 153–176.

78. — , "Joint approximations of the periods of an elliptic function by algebraic numbers," *Izv. Akad. Nauk SSSR Ser. Mat.* **22** (1958), pp. 563–576.

79. — , "An elliptic analog of an inequality of A. O. Gel′fond," *Trudy Moskov. Mat. Obšč.* **18** (1968), pp. 65–76.

80. — , "The periods of elliptic functions," *Acta Arith.* **24** (1973/74), pp. 477–489, Collection of articles dedicated to Carl Ludwig Siegel on the occasion of his seventy-fifth birthday, V.

81. — , "The algebraic independence of certain numbers," *Vestnik Moskov. Univ. Ser. I Mat. Mekh.* (1980), no. 4, pp. 46–50, 100.

82. — , "Algebraic independence of some numbers. II," *Ann. Univ. Sci. Budapest. Eötvös Sect. Math.* **25** (1982), pp. 109–123.

83. N. I. FEL′DMAN & Y. V. NESTERENKO – "Transcendental numbers," in *Number theory, IV*, Encyclopaedia Math. Sci., vol. 44, Springer, Berlin, 1998, pp. 1–345.

84. Y. Z. FLICKER – "Linear forms on arithmetic abelian varieties: ineffective bounds," *Mém. Soc. Math. France (N.S.)* **2** (1980/81), pp. 41–47, Abelian functions and transcendental numbers (Colloq., École Polytech., Palaiseau, 1979).

85. R. FRANKLIN – "The transcendence of linear forms in $\omega_1, \omega_2, \eta_1, \eta_2, 2\pi i, \log \gamma$," *Acta Arith.* **26** (1974/75), pp. 197–206.

86. É. GAUDRON – "Étude du cas rationnel de la théorie des formes linéaires de logarithmes," *J. Number Theory,* **127** (2007). no. 2, pp. 220–261.

87. — , "Formes linéaires de logarithmes effectives sur les variétés abéliennes," submitted, 2004. *Ann. Sci. de l'École Norm. Sup.,* **39** (2006), no. 5, pp. 699–773.

88. — , "Mesures d'indépendance linéaire de logarithmes dans un groupe algébrique commutatif," *Invent. Math.* **162** (2005), no. 1, pp. 137–188.

89. A. O. GEL′FOND – *Transcendental and algebraic numbers*, Translated from the first Russian edition by Leo F. Boron, Dover Publications Inc., New York, 1960.

90. P. GRAFTIEAUX – "Théorème stéphanois et méthode des pentes," in *Formes modulaires et transcendance*, Sémin. Congr., vol. 12, Soc. Math. France, Paris, 2005, pp. 179–213.

91. F. GRAMAIN – "Quelques résultats d'indépendance algébrique," in *Proceedings of the International Congress of Mathematicians, Vol. II (Berlin)*, 1998, pp. 173–182.

92. — , "Transcendance et fonctions modulaires," *J. Théor. Nombres Bordeaux* **11** (1999), no. 1, pp. 73–90, Les XXèmes Journées Arithmétiques (Limoges, 1997).

93. P. GRINSPAN – "A measure of simultaneous approximation for quasi-modular functions," *Ramanujan J.* **5** (2001), no. 1, pp. 21–45.

94. — , "Measures of simultaneous approximation for quasi-periods of abelian varieties," *J. Number Theory* **94** (2002), no. 1, pp. 136–176.
95. B. H. GROSS – "On the periods of abelian integrals and a formula of Chowla and Selberg," *Invent. Math.* **45** (1978), no. 2, pp. 193–211, With an appendix by David E. Rohrlich.
96. — , "On an identity of Chowla and Selberg," *J. Number Theory* **11** (1979), no. 3 S. Chowla Anniversary Issue, pp. 344–348.
97. T. HARASE – "The transcendence of $e^{\alpha\omega+\beta(2\pi i)}$," *J. Fac. Sci. Univ. Tokyo Sect. IA Math.* **21** (1974), pp. 279–285.
98. — , "On the linear form of transcendental numbers; $\alpha_1\omega + \alpha_3 \log \alpha_4$," *J. Fac. Sci. Univ. Tokyo Sect. IA Math.* **23** (1976), no. 3, pp. 435–452.
99. M. HINDRY & J. H. SILVERMAN – *Diophantine geometry*, Graduate Texts in Mathematics, vol. 201, Springer-Verlag, New York, 2000, An introduction.
100. N. HIRATA-KOHNO – "Formes linéaires d'intégrales elliptiques," in *Séminaire de Théorie des Nombres, Paris 1988–1989*, Progr. Math., vol. 91, Birkhäuser Boston, Boston, MA, 1990, pp. 117–140.
101. — , "Mesures de transcendance pour les quotients de périodes d'intégrales elliptiques," *Acta Arith.* **56** (1990), no. 2, pp. 111–133.
102. — , "Formes linéaires de logarithmes de points algébriques sur les groupes algébriques," *Invent. Math.* **104** (1991), no. 2, pp. 401–433.
103. — , "Nouvelles mesures de transcendance liées aux groupes algébriques commutatifs," in *Approximations diophantiennes et nombres transcendants (Luminy, 1990)*, W. de Gruyter, Berlin, 1992, pp. 165–172.
104. — , "Approximations simultanées sur les groupes algébriques commutatifs," *Compositio Math.* **86** (1993), no. 1, pp. 69–96.
105. E. M. JABBOURI – "Mesures d'indépendance algébrique de valeurs de fonctions elliptiques et abéliennes," *C. R. Acad. Sci. Paris Sér. I Math.* **303** (1986), no. 9, pp. 375–378.
106. Y. M. KHOLYAVKA – "Simultaneous approximations of the invariants of an elliptic function by algebraic numbers," in *Diophantine approximations, Part II (Russian)*, Moskov. Gos. Univ., Moscow, 1986, pp. 114–121.
107. — , "Approximation of numbers that are connected with elliptic functions," *Mat. Zametki* **47** (1990), no. 6, pp. 110–118, 160.
108. — , "Approximation of the invariants of an elliptic function," *Ukrain. Mat. Zh.* **42** (1990), no. 5, pp. 681–685.
109. — , "On the approximation of some numbers related to $\wp(z)$," *Visnik L'viv. Univ. Ser. Mekh. Mat.* **34** (1990), pp. 88–89.
110. — , "On the approximation of numbers connected with Weierstrass elliptic functions," *Sibirsk. Mat. Zh.* **32** (1991), no. 1, pp. 212–216, 224.
111. — , "Simultaneous approximation of numbers connected with elliptic functions," *Izv. Vyssh. Uchebn. Zaved. Mat.* **3** (1991), pp. 70–73.
112. — , "Zeros of polynomials of Jacobi elliptic functions," *Ukraïn. Mat. Zh.* **44** (1992), no. 11, p. 1624.
113. — , "On the approximation of certain numbers associated with Jacobi elliptic functions," in *Mathematical investigations (Ukrainian)*, Pr. L'vīv. Mat. Tov., vol. 2, L'vīv. Mat. Tov., L'viv, 1993, pp. 10–13, 106.
114. — , "On the approximation of numbers connected with $\wp(z)$," *Visnik L'vīv. Univ. Ser. Mekh. Mat.* **38** (1993), p. 64.
115. N. KOBLITZ – "Gamma function identities and elliptic differentials on Fermat curves," *Duke Math. J.* **45** (1978), no. 1, pp. 87–99.

182   Michel Waldschmidt

116. — , *Introduction to elliptic curves and modular forms*, second ed., Graduate Texts in Mathematics, vol. 97, Springer-Verlag, New York, 1993.

117. N. KOBLITZ & D. ROHRLICH – "Simple factors in the Jacobian of a Fermat curve," *Canad. J. Math.* **30** (1978), no. 6, pp. 1183–1205.

118. M. KONTSEVICH & D. ZAGIER – "Periods," in *Mathematics unlimited—2001 and beyond*, Springer, Berlin, 2001, pp. 771–808.

119. S. LANG – "Diophantine approximations on toruses," *Amer. J. Math.* **86** (1964), pp. 521–533 (= [129] pp. 313–325).

120. — , *Introduction to transcendental numbers*, Addison-Wesley Publishing Co., Reading, Mass.-London-Don Mills, Ont., 1966 (= [129] pp. 396–506).

121. — , "Transcendental numbers and Diophantine approximations," *Bull. Amer. Math. Soc.* **77** (1971), pp. 635–677 (= [130] pp. 1–43).

122. — , "Higher-dimensional Diophantine problems," *Bull. Amer. Math. Soc.* **80** (1974), pp. 779–787 (= [130] pp. 102–110).

123. — , "Diophantine approximation on abelian varieties with complex multiplication," *Advances in Math.* **17** (1975), no. 3, pp. 281–336 (= [130] pp. 113–168).

124. — , *Elliptic curves: Diophantine analysis*, Grundlehren der Mathematischen Wissenschaften, vol. 231, Springer-Verlag, Berlin, 1978.

125. — , "Relations de distributions et exemples classiques," in *Séminaire Delange-Pisot-Poitou, 19e année: 1977/78, Théorie des nombres, Fasc. 2*, Secrétariat Math., Paris, 1978, p. Exp. No. 40, 6 (= [131] pp. 59–65).

126. — , "Conjectured Diophantine estimates on elliptic curves," in *Arithmetic and geometry, Vol. I*, Progr. Math., vol. 35, Birkhäuser Boston, Boston, MA, 1983, pp. 155–171 (= [131] pp. 212–228).

127. — , *Elliptic functions*, second ed., Graduate Texts in Mathematics, vol. 112, Springer-Verlag, New York, 1987, With an appendix by J. Tate.

128. — , *Cyclotomic fields I and II*, second ed., Graduate Texts in Mathematics, vol. 121, Springer-Verlag, New York, 1990, With an appendix by Karl Rubin.

129. — , *Collected papers. Vol. I*, Springer-Verlag, New York, 2000, 1952–1970.

130. — , *Collected papers. Vol. II*, Springer-Verlag, New York, 2000, 1971–1977.

131. — , *Collected papers. Vol. III*, Springer-Verlag, New York, 2000, 1978–1990.

132. M. LAURENT – "Transcendance de périodes d'intégrales elliptiques," *C. R. Acad. Sci. Paris Sér. A-B* **288** (1979), no. 15, pp. 699–701.

133. — , "Approximation diophantienne de valeurs de la fonction Beta aux points rationnels," *Ann. Fac. Sci. Toulouse Math. (5)* **2** (1980), no. 1, pp. 53–65.

134. — , "Indépendance linéaire de valeurs de fonctions doublement quasi-périodiques," *C. R. Acad. Sci. Paris Sér. A-B* **290** (1980), no. 9, pp. A397–A399.

135. — , "Transcendance de périodes d'intégrales elliptiques," *J. reine angew. Math.* **316** (1980), pp. 122–139.

136. — , "Transcendance de périodes d'intégrales elliptiques," in *Séminaire Delange-Pisot-Poitou, 20e année: 1978/1979. Théorie des nombres, Fasc. 1*, Secrétariat Math., Paris, 1980, p. Exp. No. 13, 4.

137. — , "Transcendance de périodes d'intégrales elliptiques. II," *J. reine angew. Math.* **333** (1982), pp. 144–161.

138. K. MAHLER – "On algebraic differential equations satisfied by automorphic functions," *J. Austral. Math. Soc.* **10** (1969), pp. 445–450.

139. — , "Remarks on a paper by W. Schwarz," *J. Number Theory* **1** (1969), pp. 512–521.

140. — , "On the coefficients of the 2^n-th transformation polynomial for $j(\omega)$," *Acta Arith.* **21** (1972), pp. 89–97.

141. — , "On the coefficients of transformation polynomials for the modular function," *Bull. Austral. Math. Soc.* **10** (1974), pp. 197–218.

142. Y. MANIN – "Cyclotomic fields and modular curves," *Uspekhi Mat. Nauk* **26** (1971), no. 6, pp. 7–71, Engl. Transl. Russ. Math. Surv. **26** (1971), no. 6, pp. 7–78.

143. D. W. MASSER – "On the periods of the exponential and elliptic functions," *Proc. Cambridge Philos. Soc.* **73** (1973), pp. 339–350.

144. — , *Elliptic functions and transcendence*, Springer-Verlag, Berlin, 1975, Lecture Notes in Mathematics, Vol. 437.

145. — , "Linear forms in algebraic points of Abelian functions. I," *Math. Proc. Cambridge Philos. Soc.* **77** (1975), pp. 499–513.

146. — , "On the periods of Abelian functions in two variables," *Mathematika* **22** (1975), no. 2, pp. 97–107.

147. — , "Sur les points algébriques d'une variété abélienne," *C. R. Acad. Sci. Paris Sér. A-B* **280** (1975), pp. A11–A12.

148. — , "Transcendence and abelian functions," in *Journées Arithmétiques de Bordeaux (Conf., Univ. Bordeaux, Bordeaux, 1974)*, Soc. Math. France, Paris, 1975, pp. 177–182. Astérisque, Nos. 24–25.

149. — , "Linear forms in algebraic points of Abelian functions. II," *Math. Proc. Cambridge Philos. Soc.* **79** (1976), no. 1, pp. 55–70.

150. — , "Linear forms in algebraic points of Abelian functions. III," *Proc. London Math. Soc. (3)* **33** (1976), no. 3, pp. 549–564.

151. — , "A note on a paper of Richard Franklin: [85]," *Acta Arith.* **31** (1976), no. 2, pp. 143–152.

152. — , "Division fields of elliptic functions," *Bull. London Math. Soc.* **9** (1977), no. 1, pp. 49–53.

153. — , "A note on Abelian functions," in *Transcendence theory: advances and applications (Proc. Conf., Univ. Cambridge, Cambridge, 1976)*, Academic Press, London, 1977, pp. 145–147.

154. — , "Some vector spaces associated with two elliptic functions," in *Transcendence theory: advances and applications (Proc. Conf., Univ. Cambridge, Cambridge, 1976)*, Academic Press, London, 1977, pp. 101–119.

155. — , "The transcendence of certain quasi-periods associated with Abelian functions in two variables," *Compositio Math.* **35** (1977), no. 3, pp. 239–258.

156. — , "The transcendence of definite integrals of algebraic functions," in *Journées Arithmétiques de Caen (Univ. Caen, 1976)*, Soc. Math. France, Paris, 1977, pp. 231–238. Astérisque No. 41–42.

157. — , "Diophantine approximation and lattices with complex multiplication," *Invent. Math.* **45** (1978), no. 1, pp. 61–82.

158. — , "Heights, transcendence, and linear independence on commutative group varieties," in *Diophantine approximation (Cetraro, 2000)*, Lecture Notes in Math., vol. 1819, Springer, Berlin, 2003, pp. 1–51.

159. — , "Sharp estimates for Weierstrass elliptic functions," *J. Anal. Math.* **90** (2003), pp. 257–302.

160. D. W. MASSER & G. WÜSTHOLZ – "Algebraic independence properties of values of elliptic functions," in *Journées arithmétiques, Exeter 1980*, Lond. Math. Soc. Lect. Note Ser. 56, 360–363 , 1982.

161. — , "Fields of large transcendence degree generated by values of elliptic functions," *Invent. Math.* **72** (1983), no. 3, pp. 407–464.

162. — , "Algebraic independence of values of elliptic functions," *Math. Ann.* **276** (1986), no. 1, pp. 1–17.

163. — , "Estimating isogenies on elliptic curves," *Invent. Math.* **100** (1990), no. 1, pp. 1–24.
164. — , "Galois properties of division fields of elliptic curves," *Bull. London Math. Soc.* **25** (1993), no. 3, pp. 247–254.
165. N. D. NAGAEV – "Approximation of the transcendental quotient of two algebraic points of the function $\wp(z)$ with complex multiplication," *Mat. Zametki* **20** (1976), no. 1, pp. 47–60.
166. J. V. NESTERENKO – "On the algebraical independence of algebraic numbers to algebraic powers," in *Diophantine approximations and transcendental numbers (Luminy, 1982)*, Progr. Math., vol. 31, Birkhäuser Boston, Mass., 1983, pp. 199–220.
167. Y. V. NESTERENKO – "Algebraic independence of algebraic powers of algebraic numbers," *Mat. Sb. (N.S.)* **123(165)** (1984), no. 4, pp. 435–459.
168. — , "Degrees of transcendence of some fields that are generated by values of an exponential function," *Mat. Zametki* **46** (1989), no. 3, pp. 40–49, 127.
169. Y. V. NESTERENKO – "Measure of algebraic independence of values of an elliptic function at algebraic points," *Uspekhi Mat. Nauk* **40** (1985), no. 4(244), pp. 221–222.
170. — , "On a measure of algebraic independence of the values of elliptic functions," in *Approximations diophantiennes et nombres transcendants (Luminy, 1990)*, W. de Gruyter, Berlin, 1992, pp. 239–248.
171. — , "On the measure of algebraic independence of values of an elliptic function," *Izv. Ross. Akad. Nauk Ser. Mat.* **59** (1995), no. 4, pp. 155–178.
172. — , "Modular functions and transcendence problems," *C. R. Acad. Sci. Paris Sér. I Math.* **322** (1996), no. 10, pp. 909–914.
173. — , "Modular functions and transcendence questions," *Mat. Sb.* **187** (1996), no. 9, pp. 65–96.
174. — , "On the measure of algebraic independence of values of Ramanujan functions," *Tr. Mat. Inst. Steklova* **218** (1997), no. Anal. Teor. Chisel i Prilozh., pp. 299–334.
175. — , "Algebraic independence of π and e^{π}," in *Number theory and its applications (Ankara, 1996)*, Lecture Notes in Pure and Appl. Math., vol. 204, Dekker, New York, 1999, pp. 121–149.
176. — , "On the algebraic independence of values of Ramanujan functions," *Vestnik Moskov. Univ. Ser. I Mat. Mekh.* **2** (2001), pp. 6–10, 70.
177. Y. V. NESTERENKO & P. PHILIPPON – *Introduction to algebraic independence theory*, Lecture Notes in Mathematics, vol. 1752, Springer-Verlag, Berlin, 2001, With contributions from F. Amoroso, D. Bertrand, W. D. Brownawell, G. Diaz, M. Laurent, Yuri V. Nesterenko, K. Nishioka, Patrice Philippon, G. Rémond, D. Roy and M. Waldschmidt, edited by Nesterenko and Philippon.
178. F. PELLARIN – "The isogeny theorem and the irreducibility theorem for elliptic curves: a survey," *Rend. Sem. Mat. Univ. Politec. Torino* **53** (1995), no. 4, pp. 389–404, Number theory, II (Rome, 1995).
179. — , "The Ramanujan property and some of its connections with Diophantine geometry," *Riv. Mat. Univ. Parma (7)* **3*** (2004), pp. 275–288.
180. — , "Introduction aux formes modulaires de Hilbert et à leurs propriétés différentielles," in *Formes modulaires et transcendance*, Sémin. Congr., vol. 12, Soc. Math. France, Paris, 2005, pp. 215–269.
181. G. PHILIBERT – "Une mesure d'indépendance algébrique," *Ann. Inst. Fourier (Grenoble)* **38** (1988), no. 3, pp. 85–103.
182. P. PHILIPPON – "Indépendance algébrique de valeurs de fonctions elliptiques p-adiques," in *Proceedings of the Queen's Number Theory Conference, 1979 (Kingston, Ont., 1979)* (Kingston, Ont.), Queen's Papers in Pure and Appl. Math., vol. 54, Queen's Univ., 1980, pp. 223–235.

183. — , "Indépendance algébrique et variétés abéliennes," *C. R. Acad. Sci. Paris Sér. I Math.* **294** (1982), no. 7, pp. 257–259.

184. — , "Variétés abéliennes et indépendance algébrique. I," *Invent. Math.* **70** (1982/83), no. 3, pp. 289–318.

185. — , "Variétés abéliennes et indépendance algébrique. II. Un analogue abélien du théorème de Lindemann-Weierstrass," *Invent. Math.* **72** (1983), no. 3, pp. 389–405.

186. — , "Une approche méthodique pour la transcendance et l'indépendance algébrique de valeurs de fonctions analytiques," *J. Number Theory* **64** (1997), no. 2, pp. 291–338.

187. — , "Indépendance algébrique et K-fonctions," *J. reine angew. Math.* **497** (1998), pp. 1–15.

188. P. PHILIPPON & M. WALDSCHMIDT – "Formes linéaires de logarithmes sur les groupes algébriques commutatifs," *Illinois J. Math.* **32** (1988), no. 2, pp. 281–314.

189. — , "Formes linéaires de logarithmes elliptiques et mesures de transcendance," in *Théorie des nombres (Québec, PQ, 1987)*, W. de Gruyter, Berlin, 1989, pp. 798–805.

190. J. POPKEN & K. MAHLER – "Ein neues Prinzip Transzendenzbeweise," *Proc. Akad. Wet. Amsterdam* **38** (1935), pp. 864–871.

191. K. RAMACHANDRA – "Contributions to the theory of transcendental numbers. I, II," *Acta Arith. 14 (1967/68), 65–72; ibid.* **14** (1967/1968), pp. 73–88.

192. — , *Lectures on transcendental numbers*, The Ramanujan Institute Lecture Notes, vol. 1, The Ramanujan Institute, Madras, 1969.

193. S. RAMANUJAN – "On certain arithmetical functions," *Trans. Camb. Phil. Soc.* **22** (1916), pp. 159–184, Collected Papers of Srinivasa Ramanujan, Chelsea Publ., N.Y. 1927, N°18, 136–162.

194. G. RÉMOND & F. URFELS – "Approximation diophantienne de logarithmes elliptiques p-adiques," *J. Number Theory* **57** (1996), no. 1, pp. 133–169.

195. É. REYSSAT – "Mesures de transcendance de nombres liés aux fonctions elliptiques," in *Séminaire Delange-Pisot-Poitou, 18e année: 1976/77, Théorie des nombres, Fasc. 2*, Secrétariat Math., Paris, 1977, p. Exp. No. G22, 3.

196. — , "Mesures de transcendance de nombres liés aux fonctions exponentielles et elliptiques," *C. R. Acad. Sci. Paris Sér. A-B* **285** (1977), no. 16, pp. A977–A980.

197. — , "Travaux récents de G. V. Chudnovsky," in *Séminaire Delange-Pisot-Poitou, 18e année: 1976/77, Théorie des nombres, Fasc. 2*, Secrétariat Math., Paris, 1977, p. Exp. No. 29, 7.

198. — , "Approximation algébrique de nombres liés aux fonctions elliptiques et exponentielle," *Bull. Soc. Math. France* **108** (1980), no. 1, pp. 47–79.

199. — , "Approximation de nombres liés à la fonction sigma de Weierstrass," *Ann. Fac. Sci. Toulouse Math. (5)* **2** (1980), no. 1, pp. 79–91.

200. — , "Fonctions de Weierstrass et indépendance algébrique," *C. R. Acad. Sci. Paris Sér. A-B* **290** (1980), no. 10, pp. A439–A441.

201. — , "Propriétés d'indépendance algébrique de nombres liés aux fonctions de Weierstrass," *Acta Arith.* **41** (1982), no. 3, pp. 291–310.

202. N. SARADHA – "Transcendence measure for η/ω," *Acta Arith.* **92** (2000), no. 1, pp. 11–25.

203. T. SCHNEIDER – "Transzendenzuntersuchungen periodischer Funktionen. II. Transzendenzeigenschaften elliptischer Funktionen," *J. reine angew. Math.* **172** (1934), pp. 70–74.

204. — , "Arithmetische Untersuchungen elliptischer Integrale," *Math. Ann.* **113** (1936), pp. 1–13.

205. — , "Zur Theorie der Abelschen Funktionen und Integrale," *J. reine angew. Math.* **183** (1941), pp. 110–128.

206. — , "Ein Satz über ganzwertige Funktionen als Prinzip für Transzendenzbeweise," *Math. Ann.* **121** (1949), pp. 141–140.

207. — , *Einführung in die transzendenten Zahlen*, Springer-Verlag, Berlin, 1957.

208. J.-P. SERRE – *Cours d'arithmétique*, Collection SUP: "Le Mathématicien," vol. 2, Presses Universitaires de France, Paris, 1970, reprinted 1977. Engl. transl.: *A course in arithmetic*, Graduate Texts in Mathematics, Vol. 7. Springer-Verlag, New York, 1978.

209. S. O. SHESTAKOV – "On the measure of algebraic independence of some numbers," *Vestnik Moskov. Univ. Ser. I Mat. Mekh.* **2** (1992), pp. 8–12, 111.

210. C. L. SIEGEL – "Über die Perioden elliptischer Funktionen," *J. f. M.* **167** (1932), pp. 62–69.

211. — , *Transcendental Numbers*, Annals of Mathematics Studies, no. 16, Princeton University Press, Princeton, N. J., 1949.

212. J. H. SILVERMAN – *The arithmetic of elliptic curves*, Graduate Texts in Mathematics, vol. 106, Springer-Verlag, New York, 1986.

213. — , *Advanced topics in the arithmetic of elliptic curves*, Graduate Texts in Mathematics, vol. 151, Springer-Verlag, New York, 1994.

214. A. A. ŠMELEV – "The algebraic independence of the values of the exponential and an elliptic function," *Mat. Zametki* **20** (1976), no. 2, pp. 195–202.

215. — , "Algebraic independence of several numbers connected with exponential and elliptic functions," *Ukrain. Mat. Zh.* **33** (1981), no. 2, pp. 277–282.

216. M. TAKEUCHI – "Quantitative results of algebraic independence and Baker's method," *Acta Arith.* **119** (2005), no. 3, pp. 211–241.

217. R. TIJDEMAN – "On the algebraic independence of certain numbers," *Nederl. Akad. Wetensch. Proc. Ser. A* **74**=*Indag. Math.* **33** (1971), pp. 146–162.

218. M. TOYODA & T. YASUDA – "On the algebraic independence of certain numbers connected with the exponential and the elliptic functions," *Tokyo J. Math.* **9** (1986), no. 1, pp. 29–40.

219. R. TUBBS – "A transcendence measure for some special values of elliptic functions," *Proc. Amer. Math. Soc.* **88** (1983), no. 2, pp. 189–196.

220. — , "On the measure of algebraic independence of certain values of elliptic functions," *J. Number Theory* **23** (1986), no. 1, pp. 60–79.

221. — , "Algebraic groups and small transcendence degree. I," *J. Number Theory* **25** (1987), no. 3, pp. 279–307.

222. — , "Elliptic curves in two-dimensional abelian varieties and the algebraic independence of certain numbers," *Michigan Math. J.* **34** (1987), no. 2, pp. 173–182.

223. — , "A Diophantine problem on elliptic curves," *Trans. Amer. Math. Soc.* **309** (1988), no. 1, pp. 325–338.

224. — , "Algebraic groups and small transcendence degree. II," *J. Number Theory* **35** (1990), no. 2, pp. 109–127.

225. K. G. VASIL′EV – "On the algebraic independence of the periods of Abelian integrals," *Mat. Zametki* **60** (1996), no. 5, pp. 681–691, 799.

226. I. WAKABAYASHI – "Algebraic values of meromorphic functions on Riemann surfaces," *J. Number Theory* **25** (1987), no. 2, pp. 220–229.

227. — , "Algebraic values of functions on the unit disk," in *Prospects of mathematical science (Tokyo, 1986)*, World Sci. Publishing, Singapore, 1988, pp. 235–266.

228. — , "An extension of the Schneider-Lang theorem," in *Seminar on Diophantine Approximation (Japanese) (Yokohama, 1987)*, Sem. Math. Sci., vol. 12, Keio Univ., Yokohama, 1988, pp. 79–83.

229. M. WALDSCHMIDT – "Propriétés arithmétiques des valeurs de fonctions méromorphes algébriquement indépendantes," *Acta Arith.* **23** (1973), pp. 19–88.

230. — , *Nombres transcendants*, Springer-Verlag, Berlin, 1974, Lecture Notes in Mathematics, Vol. 402.

231. — , "Les travaux de G. V. Chudnovsky sur les nombres transcendants," in *Séminaire Bourbaki, Vol. 1975/76, 28e année, Exp. No. 488*, Springer, Berlin, 1977, pp. 274–292. Lecture Notes in Math., Vol. 567.

232. — , "Nombres transcendants et fonctions sigma de Weierstrass," *C. R. Math. Rep. Acad. Sci. Canada* **1** (1978/79), no. 2, pp. 111–114.

233. — , *Nombres transcendants et groupes algébriques*, Astérisque, vol. 69, Société Mathématique de France, Paris, 1979, With appendices by Daniel Bertrand and Jean-Pierre Serre.

234. — , "Diophantine properties of the periods of the Fermat curve," in *Number theory related to Fermat's last theorem*, Proc. Conf., Prog. Math. 26, 79–88 , 1982.

235. — , "Algebraic independence of transcendental numbers. Gel'fond's method and its developments," in *Perspectives in mathematics*, Birkhäuser, Basel, 1984, pp. 551–571.

236. — , "Algebraic independence of values of exponential and elliptic functions," *J. Indian Math. Soc. (N.S.)* **48** (1984), no. 1–4, pp. 215–228 (1986).

237. — , "Groupes algébriques et grands degrés de transcendance," *Acta Math.* **156** (1986), no. 3–4, pp. 253–302, With an appendix by J. Fresnel.

238. — , "Some transcendental aspects of Ramanujan's work," in *Proceedings of the Ramanujan Centennial International Conference (Annamalainagar, 1987)*, RMS Publ., vol. 1, Ramanujan Math. Soc., 1988, pp. 67–76.

239. — , "Sur la nature arithmétique des valeurs de fonctions modulaires," *Astérisque* **245** (1997), p. Exp. No. 824, 3, 105–140, Séminaire Bourbaki, Vol. 1996/97.

240. — , "Density measure of rational points on abelian varieties," *Nagoya Math. J.* **155** (1999), pp. 27–53.

241. — , "Transcendance et indépendance algébrique de valeurs de fonctions modulaires," in *Number theory (Ottawa, ON, 1996)*, CRM Proc. Lecture Notes, vol. 19, Amer. Math. Soc., Providence, RI, 1999, pp. 353–375.

242. — , "Algebraic independence of transcendental numbers: a survey," in *Number theory*, Trends Math., Birkhäuser and Hindustan Book Agency, Basel and New-Delhi, 2000, pp. 497–527.

243. — , "Transcendence of periods: the state of the art," *Pure and Applied Mathematics Quarterly* **2** (2006), no. 2, pp. 199–227.

244. E. WHITTAKER & G. WATSON – *A course of modern analysis. An introduction to the general theory on infinite processes and of analytic functions; with an account of the principal transcendental functions*. 4th ed., reprinted, Cambridge: At the University Press. 608 p., 1962.

245. J. WOLFART & G. WÜSTHOLZ – "Der Überlagerungsradius gewisser algebraischer Kurven und die Werte der Betafunktion an rationalen Stellen," *Math. Ann.* **273** (1985), no. 1, pp. 1–15.

246. G. WÜSTHOLZ – "Algebraische Unabhängigkeit von Werten von Funktionen, die gewissen Differentialgleichungen genügen," *J. reine angew. Math.* **317** (1980), pp. 102–119.

247. — , "Sur l'analogue abélien du théorème de Lindemann," *C. R. Acad. Sci. Paris Sér. I Math.* **295** (1982), no. 2, pp. 35–37.

248. — , "Über das Abelsche Analogon des Lindemannschen Satzes. I," *Invent. Math.* **72** (1983), no. 3, pp. 363–388.

249. — , "Recent progress in transcendence theory," in *Number theory, Noordwijkerhout 1983*, Lecture Notes in Math., vol. 1068, Springer, Berlin, 1984, pp. 280–296.

250. — , "Transzendenzeigenschaften von Perioden elliptischer Integrale," *J. reine angew. Math.* **354** (1984), pp. 164–174.
251. — , "Algebraische Punkte auf analytischen Untergruppen algebraischer Gruppen," *Ann. of Math. (2)* **129** (1989), no. 3, pp. 501–517.
252. A. Y. YANCHENKO – "On the measure of algebraic independence of values of derivatives of a modular function (the p-adic case)," *Mat. Zametki* **61** (1997), no. 3, pp. 431–440, Engl. Transl. Math. Notes **61**, 3 (1997), 352–359.
253. K. R. YU – "Linear forms in elliptic logarithms," *J. Number Theory* **20** (1985), no. 1, pp. 1–69.
254. D. ZAGIER – "Introduction to modular forms," in *From Number Theory to Physics (Les Houches, 1989)*, Springer, Berlin, 1992, pp. 238–291.